The
FUTURE
of
EAST ASIA

Remarks on Regional Cooperation in Asia

LIU Zhenmin

The Future of East Asia:
Remarks on Regional Cooperation in Asia

Author:	Liu Zhenmin
Editor:	Wu Shicun
Cover Design:	Cathy Chiu
Publisher:	The Commercial Press (H.K.) Ltd.
	8/F, Eastern Central Plaza, 3 Yiu Hing Road,
	Shau Kei Wan, Hong Kong
	http://www.commercialpress.com.hk
Printing:	Elegance Printing & Book Binding Co. Ltd.,
	Block A, 4/F, Hoi Bun Industrial Building, 6 Wing
	Yip Street, Kwun Tong, H.K.
	©2020 The Commercial Press (H.K.) Ltd.
	First edition, First printing, March 2020
	ISBN 978 962 07 6638 1
	Printed in Hong Kong

Contents

Editor's Note

In 2010, China overtook Japan to become Asia's largest and the world's second largest economy. The United States subsequently adopted what it called a strategy of "re-balance" to Asia and the Pacific. The move was seen as an unprecedented test and challenge to China's neighborhood foreign policy, and raised the question of where the relations between China and East Asian nations would be heading and how such relationship would evolve. Given the developments, stronger Asian regional cooperation was believed to be a viable way forward.

Since the 18th National Congress of the Communist Party of China (CPC) held in November 2012, China has made new strides and broken new grounds in strengthening relations with its neighboring countries under the wise leadership of the CPC Central Committee with Comrade Xi Jinping at its core. China's neighborhood diplomacy has entered a new era. In 2013, the CPC Central Committee held a work conference exclusively on China's relations with neighbors and its neighborhood policy. It was the first meeting of its kind since the founding of the People's Republic. The meeting established the guiding principles, strategic objectives and general tasks for China's neighborhood policy in the new era.

In the wake of the conference, China has developed a new thinking on foreign policy and launched a number of new diplomatic initiatives, such as building a new model of international relations centering on win-win cooperation, taking an approach of amity, sincerity, mutual benefit and inclusiveness to its neighbors, and launching a new vision for Asian security. The new initiatives also include building a China-ASEAN and Asian community of shared future and a Silk Road Economic Belt and a 21st Century Maritime Silk Road (Belt and Road Initiative or BRI), and an Asian Infrastructure Investment Bank (AIIB). The new thinking and new initiatives represent part and parcel of China's foreign and neighborhood policy in the new era.

Guided by the new thinking and based on its new initiatives, China as taken actions on the ground to promote and deepen regional cooperation in Asia. They include:

- Initiating a "2+7 cooperation framework" for China-ASEAN cooperation. The "2" means the "two wheels" that drive the growth of China-ASEAN relations: one is cooperation in the political and security fields; the other is economic development. The "7" refers to the seven priority areas of cooperation, i.e., good-neighborly friendship, trade and investment, financial cooperation, connectivity, maritime cooperation, and cultural and people-to-people exchanges.
- Promoting a closer ASEAN plus China, Japan and the ROK cooperation (10+3 cooperation) and building an East Asia economic community as the ultimate goal of East Asian cooperation.
- Leveraging the constructive role of the East Asia Summit and ASEAN Regional Forum in East Asian cooperation.
- Advancing Asian security cooperation, and building a China-led dialogue platform on security; and
- Seeking a proper settlement of regional hotspot issues.

Mr. Liu Zhenmin was appointed assistant foreign minister in early 2010 upon completion of his overseas assignment as Ambassador Extraordinary & Plenipotentiary and Chinese Deputy Permanent Representative to the United Nations. As of 2011, his portfolio as assistant minister covered Asia. In February 2012, Liu was named Ambassador Extraordinary & Plenipotentiary and Chinese Permanent Representative to the United Nations Office at Geneva. In May 2013, he was recalled from Geneva and appointed Vice Minister for Foreign Affairs with a portfolio covering Asia among others. He served as Chinese Vice Foreign Minister until late July 2017 when he started to serve as United Nations Under-Secretary-General for Economic and Social Affairs.

As a witness and participant of Asia's regional cooperation process in recent years, Mr. Liu Zhenmin has systematic and in-depth observations on Asian regional cooperation and how to promote such cooperation.

During his tenure as Vice Foreign Minister responsible for Asian affairs, Liu Zhenmin acted on the guiding principles established at the 18th CPC Congress and General Secretary Xi Jinping's new vision, new thinking, and new strategy for the governance of China. He used all opportunities and occasions to articulate China's position and vision and make China's voice heard, presenting a real China to the world and safeguarding China's interests. He also made a commendable effort to promote Asian regional cooperation and an Asian community with a shared future.

With Liu's consent, we have compiled this collection of his public essays and speeches in recent years for the benefit of readers.

This collection includes 34 essays on Asian regional cooperation that Liu Zhenmin wrote from August 2011 to July 2017. They are in the form of thematic papers, commentaries or public speeches. The essays focus on China's foreign policy and diplomatic initiatives, such as the Belt and Road Initiative, China-ASEAN relations, and cooperation among the South China Sea coastal states, and also cover regional hotspot issues. They provide a systematic briefing on China's vision, approach and policy in respect of Asian regional cooperation, and come to an unambiguous conclusion that the future of East Asia lies in stronger and closer regional cooperation.

I wish to draw readers' special attention to his essay entitled "Promoting East Asian Regional Cooperation in Response to Changes in Global Landscape". It is a thematic paper completed by Liu Zhenmin on his own during his training at a Central Party School program for officials at or above the level of vice minister and vice provincial governor from June to August 2011. In his paper, the author has, for the first time, made a systematic analysis and observation of his own of the features and difficulties of East Asian cooperation and proposed a way forward for the future of East Asia.

Asian regional cooperation is a new topic, and promoting Asian regional cooperation will be a long-term mission in China's foreign affairs. Hopefully, the publication of this collection will be of some help to government officials and academics involved in Asian regional cooperation, and other readers who take an interest in Asian affairs and studies, and will facilitate their research and studies of the issue.

Wu Shicun
President of the National Institute for South China Sea Studies
Chairman of the Board of Directors of the China-Southeast
Asia Research Center on the South China Sea
August 2018

Promoting East Asian Regional Cooperation in Response to Changes in Global Landscape[1]

August 2011

Since the start of the 21st century, especially the outbreak of the international financial and economic crisis, the world economic and political landscape has undergone significant changes. The world's economic center of gravity is shifting eastward at a faster pace and the strategic position of the Asia-Pacific is becoming increasingly important. A most prominent change is the fact that East Asian nations are among the first to have recovered from the crisis and China's economic strength and impact have risen markedly, and that there has arisen a rare opportunity for economic growth for countries in East Asia.

In contrast to such a good momentum, regional cooperation in East Asia has been moving at a snail's pace. Is it possible for East Asia to become a third pole of the world economy next only to the EU and the North American Free Trade Area (NAFTA)? That will depend on how well East Asian regional cooperation will develop. The rise of East Asia has not come easily, and the key to a stronger East Asia lies in its unity. That is the only path for East Asia to gain strength.

With China overtaking Japan as the world's second largest economy in 2010, the world is keeping a close watch on China's development, its attitude toward East Asian cooperation and its role in such cooperation. A prevailing view is that China will become a global power after a few more decades of development.

From a historical perspective, all global powers have invariably been regional powers in the first place. In the case of China, given its geographical location, it needs to be a major player in Asia before taking

---

1   This is a paper written by Liu Zhenmin during his training at the 49th edition of the Central Party School training program for officials at or above the level of vice minister and vice provincial governor. The paper was published in August 2011 in the School's collection of research reports on the thematic topic of "Changes in World Economic and Political Landscape and Possible Measures in Response".

on any global status. To become a major player in its region, China must work together with other East Asian nations to make regional cooperation a success. China needs to have an Asia-Pacific and global vision while focusing on East Asia and take actions on the ground to promote regional cooperation in East Asia.

I History and current status of regional cooperation in East Asia

1. Geopolitical background of regional cooperation in East Asia

In geographical terms, East Asia includes Northeast Asia and Southeast Asia. In Northeast Asia, there was no regional cooperation of any form before the end of the Cold War. In Southeast Asia, an Association of Southeast Asian Nations (ASEAN) was formed in the 1960s with a focus on political and security cooperation. It was not until the end of the Cold War did ASEAN shift its focus to economic cooperation. In 1992, the six ASEAN founding members decided to establish an ASEAN free trade area (FTA) in 15 years' time as from 1993. Its purpose was to enhance ASEAN members' competitiveness in the world economy. In 2008, the FTA of ASEAN was up and running. ASEAN now has ten member countries, all from Southeast Asia, and has established an FTA of its own. ASEAN has developed into a major sub-regional organization in East Asia. In the meantime, due to difficulty in coordination between China and Japan, both being the region's major powers, ASEAN has since its inception automatically played the role of a "facilitator" of regional cooperation in East Asia.

In geopolitical terms, some East Asian countries, such as Japan, the ROK, and the Philippines, have been allies of the United States since the end of World War II. The United State has wielded a considerable influence in East Asia. Russia, with its vast territory in Asia, pays close attention to East Asia's regional cooperation. Developed Pacific countries like Canada, Australia and New Zealand also take a strong interest in East Asia. As China, Japan and the ROK have a growing weight in the world economy, the United States and other countries have intensified efforts to build up their influence in East Asia through the expansion of Asia-Pacific cooperation.

In terms of international relations, East Asian nations were members of either the Eastern or Western bloc during the Cold War. The bipolar world in the wake of World War II had a huge impact on East Asian nations. The Korean Peninsula is still in Cold War mode. All this made it impossible for East Asia to develop any regional cooperation before the end of the Cold War.

2. From Asia-Pacific cooperation to East Asian cooperation

In the late 1980s, some Asia-Pacific countries, having seen the rise of the European Community and the birth of the NAFTA, broached the idea of an Asia-Pacific regional cooperation mechanism. In November 1989, thanks to the initiative and efforts of Australia and Japan, the Asia-Pacific Economic Cooperation (APEC) came into being and was open to economies across the Pacific. In the first few years following its inception, APEC only had annual meetings of ministers responsible for foreign affairs and trade, which primarily considered issues of trade and investment liberalization, business facilitation and economic and technological cooperation in the Asia-Pacific region.

In 1993, the first APEC Economic Leaders' Meeting was held at then US President Bill Clinton's initiative. APEC economic leaders have since met every year, leading to a sharp rise of influence of APEC in the world.

In 1994, then Indonesian President Suharto initiated the ambitious Bogor Goals, which were identified and endorsed by APEC members as APEC's major pursuit in the years to come. The Bogor Goals also had a strong repercussion in Europe.

To achieve the Bogor Goals, APEC came up with an Osaka Action Agenda in 1995 and a Manila Action Plan in 1996. APEC was then widely seen as a champion of Asia-Pacific trade and investment liberalization, and its global influence increased tremendously. Unfortunately, however, APEC, having grown for eight years, lost its momentum and started to ebb from 1997 onward.

The Asian financial crisis in 1997 had a huge adverse impact on APEC, taking tolls on many of its member economies. Indonesia, which initiated the Bogor Goals, was overwhelmed by the crisis and could no longer spare efforts for trade liberalization. APEC, however, did little vis-à-vis the onslaught of the crisis. Still more disappointing was the fact that APEC's developed economies even agreed to the

International Monetary Fund (IMF) imposing harsh restrictions on the crisis-stricken Asian members. This has led to the loss of confidence in APEC among ASEAN members. In sharp contrast to some APEC members' indifference, China took prompt actions to prevent further escalation of the Asian financial crisis and announced immediately that RMB would not depreciate. China's announcement was welcomed by ASEAN countries. It was in the same year that ASEAN started its East Asian cooperation process, which resulted in the birth of a 10+3 (i.e., ASEAN plus China, Japan, ROK) cooperation framework and three 10+1 cooperation frameworks.

3. Coordination and competition between East Asian cooperation and Asia-Pacific cooperation

The establishment of the 10+3 (ASEAN plus Three) and 10+1 cooperation frameworks marked the official launch of East Asian regional cooperation. China firmly supports the 10+3 and 10+1 cooperation. Soon after the outbreak of the Asian financial crisis, Japan indicated that it would contribute US$100 billion for the establishment of an Asian Monetary Fund to help East Asian countries tide over difficulties, but its announced contribution was not made available due to the US opposition. The ROK has vigorously advocated the vision of an East Asian Community. In 1998, then ROK President Kim Dae-jung proposed the establishment of an East Asia Vision Group (EAVG). At the 10+3 Leaders' Meeting in 2000, Kim Dae-jung again proposed the launch of an EAVG. Economic and political academics, especially those from ROK, China and Japan, have conducted studies on the issue.

However, due to geopolitical reasons, the development of regional cooperation in East Asia has, from its outset, been in the shadow of Asia-Pacific cooperation, and under the sway of the United States. As a result, East Asian cooperation and Asia-Pacific cooperation have over the past 15 years developed in parallel on two separate tracks amidst coordination and competition. Thanks to all parties' efforts over the years, there have been formed a number of frameworks and negotiation platforms as indicated below for Asia-Pacific cooperation and East Asian cooperation respectively.

(1) Asia-Pacific cooperation

A. APEC is the main channel of Asia-Pacific cooperation. It was established in 1989 and has 21 member economies (including Hong Kong and Chinese Taipei as non-sovereign economies). With a population of nearly 2.7 billion and a GDP accounting for more than half of the world's total, APEC carries considerable weight in the global economy. APEC members include China, Japan, the ROK, seven ASEAN members, the United States, Canada, Australia, New Zealand and Russia. Over the past 22 years, APEC has made some progress in trade and investment liberalization and facilitation, regional economic integration and economic and technological cooperation. The establishment of a Free Trade Area of the Asia-Pacific (FTAAP) has been identified as its long-term goal. That said, APEC remains an economic cooperation forum in nature.

B. The Trans-Pacific Strategic Economic Partnership Agreement (TPP) is a multilateral free trade arrangement initiated by Singapore, Brunei, New Zealand and Chile in 2006. In November 2009, with US high-profile participation in TPP negotiations, the TPP influence grew rapidly. As of the end of 2010, Australia, Malaysia, Peru and Vietnam joined in successively, and the number of participating countries increased to nine. In their meeting in 2010, APEC Economic Leaders agreed that TPP could be a way forward for achieving the ultimate goal of an FTAAP. The United States sees the TPP as a "fast track" for reshaping the Asia-Pacific integration process and for harmonizing the envisaged Asia-Pacific FTA in accordance with its own will and standards. To this end, the United States hopes to have an agreement before the 2011 APEC meeting in Hawaii.

(2) East Asian cooperation

A. 10+3 cooperation, three 10+1 cooperation frameworks, trilateral dialogue and cooperation among China, Japan and the ROK.

The 10+3 Leaders' Meeting is a fully-operational framework for dialogue and cooperation in East Asia. In 1997, thanks to the efforts of Malaysia, the host of the 2nd ASEAN Summit, an informal meeting of ASEAN, Chinese, Japanese and ROK leaders was made possible. Since then, ASEAN, Chinese, Japanese and ROK leaders have met every year with ASEAN as the host. Since 1999, the 10+3 cooperation has entered a stage where all parties have taken a greater interest in practical and results-oriented cooperation. While working toward the establishment of a 10+3 free trade area, the 10+3 leaders now pay more attention to cooperation on the ground in the economic, social, political and other fields. To this end, a Chiang Mai Initiative for Financial Cooperation and a Rice Emergency Reserve Initiative for Food Security have been adopted. The 10+3 leaders support connectivity between ASEAN countries. Forums on East Asian cooperation have been hosted by ASEAN countries, China, Japan and the ROK alternately. The APEC Economic Leaders' Meeting in 2010 agreed that a 10+3 Free Trade Area would help facilitate the realization of an FTAAP.

The three 10+1 collaboration frameworks complement the 10+3 cooperation. Following the launch of the 10+3 dialogue in 1997, ASEAN held separate summit meetings with China, Japan and the ROK. This has been established as a framework for 10+1 dialogue and cooperation. ASEAN has established three separate free trade areas with China, Japan and the ROK.

In 2002, China and ASEAN signed a Comprehensive Economic Cooperation Framework Agreement, and started FTA negotiations. On January 1, 2010, a China-ASEAN FTA was up and running. The FTA covers an area of 14 million square kilometers, with a population of 1.9 billion, a gross domestic product of more than US$60 billion and a total foreign trade of more than US$450 million. It is the first FTA that China has established with a foreign entity, so is the first that ASEAN has built with a non-ASEAN country. In 2010, China's trade with ASEAN reached an all-time high of US$279.78 billion, an increase of 37.5 percent over the previous year. China is ASEAN's largest trading partner as compared to Japan and the ROK.

- (b) Japan and ASEAN started FTA negotiations in 2005 and reached an FTA agreement in 2008.
- (c) The ROK and ASEAN launched FTA negotiations in 2003, and reached relevant agreements in 2006 and 2007.

China-Japan-ROK (CJK) cooperation is the key to the success of the 10+3 cooperation and regional cooperation in East Asia. Since 1999, the leaders of the three countries have held informal meetings on a regular basis. The three countries have established ministerial, senior official and working-level meeting mechanisms in many fields, and have carried out fruitful cooperation in the fields of economy and trade, information, environmental protection, human resources development and culture. The trilateral summit held this year agreed to complete the study on the trilateral FTA and a decision would be made next year on the start of the negotiations.

B. East Asia Summit (EAS)

The first EAS meeting was held in Kuala Lumpur on December 14, 2005. The ASEAN member countries, China, Japan, the ROK, Australia, New Zealand and India were represented at the meeting. The idea of an East Asia summit was initiated by Malaysian Prime Minister Mahathir in 2000. Its purpose is to build on the 10+3 dialogue mechanism and develop a higher level cooperation mechanism and work toward an East Asian Economic Community. The United States strongly opposed the idea for fear that the EAS would weaken APEC's role and diminish US influence in East Asia. In 2002, 10+3 leaders agreed at their summit that "promoting the evolution of the 10+3 Leaders' Meeting to the East Asia Summit" would be one of the medium and long-term goals of 10+3 cooperation. In 2004, following a heated debate, the 10+3 countries decided to create a new platform known as East Asia Summit while maintaining the 10+3 framework. It was also agreed that Australia, New Zealand and India would be invited to attend the EAS. In 2010, with the formal participation of the United States and Russia, the EAS membership increased to 18 countries.

The EAS is a strategic forum participated and guided by leaders. It has become an annual event hosted by ASEAN members. Its primary function is to have dialogues on strategic, political and economic issues of common interest and concern. The EAS has issued statements or declarations on cooperation in bird flu prevention and control, energy security, climate change, global economic and financial crisis, disaster management, and reconstruction of Nalanda University. The study of an East Asia Comprehensive Economic Partnership (CEPEA) under the EAS is progressing smoothly. The 4th EAS agreed to advance the CEPEA study in parallel with that of an East Asia Free Trade Area (EAFTA) within the 10+3 framework. ASEAN has established a working group for the purpose.

C. ASEAN Regional Forum (ARF)

In 1994, ASEAN established an ASEAN Regional Forum (ARF) as a move of preventive diplomacy. The ARF has 27 members, namely, 11 Southeast Asian nations (10 ASEAN members and East Timor); four South Asian nations (India, Pakistan, Bangladesh and Sri Lanka); five Northeast Asian countries (China, Japan, the ROK, the DPRK and Mongolia); three South Pacific countries (Australia, New Zealand and Papua New Guinea); two from North America (United States and Canada); and Russia and the European Union. According to the ARF Concept Paper, the ARF process is divided into three stages: confidence-building, preventive diplomacy, and conflict resolution. Preventive diplomacy is a key issue throughout the forum and the focus of all the parties. The ARF has annual meetings of foreign ministers and senior officials, which are hosted by ASEAN countries. The ARF is a platform for all parties to discuss regional hotspot issues.

In addition, there are two sub-regional economic cooperation projects in East Asia. One is the Greater Mekong Sub-region Program. The program was initiated by the Asian Development Bank and implemented in 1992. Participants all come from the Mekong River Basin, including Laos, Myanmar, Cambodia, Thailand, Vietnam and Yunnan Province of China. The cooperation covers seven areas: transportation, energy, telecommunications, the environment, tourism, human resource development and trade and investment. In 2005, Guangxi Zhuang Autonomous Region of China joined the mechanism and initiated a vision of "One Axis, Two Wings" for China-ASEAN regional economic cooperation. Hainan Province of China is taking necessary steps to join the mechanism, hoping to work with Vietnam's eastern ports to develop logistics service-led cooperation in the pan-Beibu Gulf area. Through such cooperation, Hainan seeks to ride on the "two wings" and contribute to the development of the Yangpu Bonded Port Area and make Hainan an attractive destination of international tourism.

The other is the common development of the Greater Tumen River. Initiated and launched by the United Nations Development Program in October 1991, the project is a platform for economic exchanges and cooperation in Northeast Asia and participated by China, the DPRK, the ROK, Russia and Mongolia. China has designated Yanbian Prefecture, Changchun City and Jilin City in Jilin Province as areas for international cooperation in the development of Tumen River.

II Features and difficulties of regional cooperation in East Asia

1. Features of East Asian regional cooperation

From the development and evolution of the above-mentioned regional cooperation in East Asia, one can see that regional cooperation in East Asia has just started and lacks a full-fledged operational mechanism. There is still a long way to go before an East Asian regional organization can be up and running. Compared with other regions, East Asian regional cooperation has features of its own.

First, there are huge differences among members. The most prominent is their uneven level of economic development. East Asia includes developed countries such as Japan, a large number of developing countries, as well as several least developed countries, which makes it more difficult to negotiate trade liberalization and economic integration. Some countries are big while others are small, all having different concerns of their own. Small countries may worry they will be marginalized by economic globalization and regional integration, while big countries may be afraid of losing their competitive advantages and status that they have long enjoyed. In terms of political system, there are capitalist and socialist countries. Their views on regional hotspot issues vary greatly. In terms of religion and culture, there are Taoism, Buddhism, Islam, Shinto, Christianity and other faiths. Such tremendous differences in so many aspects make it difficult for East Asian nations to deepen regional cooperation.

Second, the operational modality of East Asian cooperation has been primarily informal. Although the word "informal" was removed from the 10+3 Leaders' Meeting in 2000, the EAS and the ARF meetings still have the prefix of "informal". The decision-making of the three frameworks is a consensus-based process.

Third, the cooperation mechanisms are still in the form of a forum in nature. The EAS is a strategic forum participated by state leaders and the ARF is a forum at the level of foreign ministers. The 10+3 cooperation is primarily in the form of dialogue or talks. If there are any deep-rooted issues in cooperation to be addressed, a separate process of negotiation has to be started.

Fourth, the operation of cooperation mechanism is led by ASEAN.

All meetings within the 10+3 cooperation framework are hosted by rotating ASEAN chair; the EAS is hosted by ASEAN countries; and the ARF is after all an ASEAN forum, and other countries participate at ASEAN's invitation.

Fifth, its membership overlaps with APEC. Apart from the 10+3 cooperation, the EAS and ARF include most APEC members, especially its major members like China, Japan and the United States.

Sixth, developed members always seek to hold sway or influence the agenda and discussion of the EAS and the ARF.

Seventh, hotspot issues in East Asia remain of concern. In addition to disputes over islands and maritime issues between countries concerned, some EAS and ARF countries keep hyping up the denuclearization of the Korean Peninsula. Some countries always attempt to raise the South China Sea issue, the Myanmar issue and the Fiji issue, which undoubtedly hinders the build-up of political mutual trust and affects the discussion on regional cooperation.

2. Difficulties in East Asian regional cooperation

The 15-year history of East Asian cooperation shows that every step of its cooperation has been difficult. This is due to the composition of countries of East Asian cooperation and the location of these nations in the Asia-Pacific region. Generally speaking, the major difficulty of East Asian cooperation comes from its relationship with Asia-Pacific cooperation, especially with the United States, and from the issue of which country should lead East Asian cooperation. The reasons for such difficulties are as follows.

First, ASEAN seeks to lead the process of East Asian regional cooperation, but faces growing challenges. First of all, maintaining "balance of power" in the region has always been the strategy of some ASEAN countries. They welcome more investment from the United States, Australia in East Asia to balance the impact of China and Japan. Secondly, they worry that the accession of big powers will further weaken ASEAN's unique leading role and may result in its marginalization. In fact, in the past few years, in the process of East Asian cooperation, ASEAN was seen as "a small horse pulling a big cart" and such pressures still remain.

Second, Japan is ambivalent about East Asian regional cooperation and its policy is inconsistent. Japan is a major power in East Asia. Until 2009, Japan had been Asia's largest economy and played an important part in East Asia's economic development. But Japan has been playing a tactics of balance between East Asia and the United States. At the outbreak of the Asian financial crisis, Japan was quite enthusiastic about the 10+3 cooperation. It once indicated that it would contribute US$100 billion to establish an Asian Monetary Fund to help East Asian countries tide over hard times, but its promise has never come true due to the US opposition. However, with China's increasing influence in East Asia, especially on ASEAN, Japan began to question the 10+3 cooperation. In 2002, Japanese Prime Minister Junichiro Koizumi proposed to include Australia, New Zealand and India in the 10+3 cooperation to establish an "East Asian community". His idea was supported by Singapore and some other ASEAN members. This has led to the creation of an EAS (10+6) in 2005 while the 10+3 cooperation mechanism remains in place.

In September 2009, the Democratic Party of Japan (DPJ) took office and pursued a foreign policy of "returning from the United States to Asia". At his meeting with Chinese President Hu Jintao in New York and in his address to the United Nations General Assembly, Prime Minister Hatoyama put forward the idea of building an East Asian community on the basis of 10+3. At the meeting of leaders of China, Japan and the ROK held on October 10, 2009, an "East Asian community" was included in the Joint Statement on the Tenth Anniversary of Trilateral Cooperation among China, Japan and the ROK as a long-term goal. At the EAS in Thailand in late October, Prime Minister Hatoyama elaborated on this idea and sought the support of ASEAN countries. However, after the government of Naoto Kan taking office in 2010, Japan stopped its advocacy of an East Asian community, and highlighted the role of APEC instead. It publicly stated that Japan would consider joining the TPP negotiations. At present, Japan claims to be supportive of the 10+3 cooperation, but it is not as enthusiastic as before.

Third, the United States attempts to weaken and offset East Asian regional cooperation and East Asian integration by enhancing Asia-Pacific and trans-Pacific economic cooperation. With China's growing influence in East Asia and the growing momentum of East Asian cooperation with the 10+1 and 10+3 at its core, the post-WWII US dominance in Asia-Pacific affairs through collaboration with its allies and partners can no longer be sustained. The United States has always been opposed to any idea of East Asian regional cooperation that would exclude itself. The United States backs stronger trans-Pacific arrangements such as APEC and TPP, and hopes to maintain its swaying power in East Asian affairs and increase its influence over China, India and ASEAN in order to remain dominant in Asia-Pacific affairs.

The United States has indicated recently that strengthening multilateral mechanisms will be the top priority of its Asia-Pacific policy in the coming years. It hopes that ASEAN, APEC, EAS, ARF, TPP, Pacific Islands Forum and other frameworks will all play an effective role. The United States considers the EAS more important than 10+3 for it is more open and inclusive. At present, the TPP negotiations have made positive progress, and the United States will press ahead with the relevant negotiations.

Fourth, Australia has been overtly and covertly opposed to the idea of 10+3 cooperation and an East Asian community. In 2008, Prime Minister Kevin Rudd proposed an initiative to establish the Asia-Pacific community. On December 4, 2009, Australia hosted in Sydney a meeting on Asia-Pacific community in response to the idea of an East Asian community proposed by Japanese Prime Minister Hatoyama and recognized by the China-Japan-ROK Trilateral Summit. In his speech, Prime Minister Kevin Rudd repeatedly stressed the need for an Asia-Pacific community to include all the major powers, especially the United States. He emphasized that the Asia-Pacific community should either be established as a new institution, or established on the basis of APEC, the EAS and the ARF. The purpose of Australia is to share the development dividends of East Asia and enhance its position in the Asia-Pacific region by playing the role of "balancer" and "communicator" between East Asia and the United States.

III Views and suggestions on how to promote regional cooperation in East Asia

1. Views

Economic globalization and regional economic integration are two major trends in today's world economic development. The mutual complementarity of East Asian economies provides favorable conditions for economic integration in East Asia. Driven by the economic globalization, East Asian economies have become an integral part of the globalized production network. The economic links between them are getting closer and stronger. Intra-regional trade is developing rapidly and trade complementarity is increasing. The acceleration of globalization has promoted the transnational flow of capital, technology, information, resources, labor and other factors. In order to reduce the risks brought by such international flows, it is particularly important for East Asia to strengthen cooperation. In fact, both the Asian financial crisis in 1997 and the international financial and economic crisis since 2008 have shown that APEC has done little to overcome or alleviate difficulties in the East Asian economic development, and is in fact unable to do so. In order to promote better and faster growth of the East Asian economy in the post-crisis era, and to cope with the new round of international economic competition, East Asian countries should unite to strengthen themselves and take greater steps on the road of regional cooperation. In the process of promoting East Asian cooperation, we should keep the following points in mind:

(1) East Asian cooperation should be open, inclusive and mutually beneficial. To promote East Asian cooperation, we should properly handle its relations with Asia-Pacific cooperation and with countries outside the region. Any modality to be adopted for East Asian cooperation must be based on East Asia's reality. In the case of 10+3 cooperation, we may consider a gradual and step-by-step approach and take it first as sub-regional cooperation under Asia-Pacific cooperation. The 10+3 cooperation may continue as the main avenue for East Asian cooperation and this is extremely important for East Asian cooperation to achieve quick results on the ground. Efforts should be made to enhance trust and dispel suspicion and mistrust harbored by countries outside the region. We need to highlight the fact all countries in East Asia value both Asia-Pacific cooperation and East Asian cooperation. The 10+3 and APEC may develop parallel and both mechanisms should avoid procedural exclusivity. For cooperation mechanisms open to countries outside the region, the principle of inclusiveness is essential. It is under such principle that we welcome the United States and Russia to join the EAS, and keep the door open to other countries.

(2) East Asian cooperation must proceed in a gradual and orderly manner and always be results-oriented. It must be based on East Asia's defining feature of diversity, and start with easy things and take on

the difficult at a later stage. Such approach can help enhance mutual complementarity and reinforcement of the existing frameworks of cooperation. At this stage, the focus will be placed on the buildup of the existing cooperation mechanisms, and on the envisaged East Asia FTA, financial cooperation and infrastructure connectivity through the 10+1, 10+3, CJK cooperation frameworks. The conditions for political and security cooperation between East Asian countries is still not mature. Therefore, the EAS and the ARF may remain a forum in nature with a view to establishing political mutual trust. Both forums should refrain from introducing any controversial issues on their agenda.

(3) East Asian cooperation must remain committed to ASEAN's leading role. ASEAN's leading role in East Asian cooperation is formed over the years. It was ASEAN that initiated cooperation mechanisms, such as informal meetings of 10+3 and 10+1 leaders, in the early stage of East Asian integration. The meetings have made it possible for East Asian leaders to sit together and exchange views on regional cooperation. Continued leading role of ASEAN in the existing cooperation mechanisms not only represents ASEAN's desire, but also conforms to the mentality of big powers concerned. First, ASEAN is most interested in and enthusiastic about stronger cooperation in East Asia. ASEAN does not want to be dominated by big powers. It fears that its leading position will have to be conceded to others. ASEAN is willing to continue its leading and mediator's role. Second, the political environment in the region makes it impossible and impermissible for China, Japan or the ROK to exclusively lead the process of East Asian cooperation. The United States for its part hates to see China or Japan take the lead in East Asian cooperation. Although ASEAN may sometimes find itself unable to adequately coordinate relations among major powers as it hopes, its leading role in the East Asian cooperation process remains a practical necessity and is acceptable to all parties concerned.

The role of China, Japan and the ROK should be enhanced in East Asian cooperation. The 10+3 FTA plays an irreplaceable role in deepening East Asian cooperation. ASEAN has established separate FTAs with China, Japan and the ROK. A CJK FTA is the prerequisite to an FTA of ASEAN and CJK. The CJK trade and GDP account for 75 percent and 90 percent of that of East Asia respectively. CJK cooperation is crucial to East Asia's stability and prosperity. At present, the building of CJK FTA has entered a critical stage. Since the joint study of FTA was launched one year ago, consensus among the three countries on issues concerned has been growing, and the conditions for formal launch of the FTA negotiations are maturing. The United States is pressing ahead with TPP negotiations in an attempt to dominate the process of East Asia and Asia-Pacific economic cooperation. However, with the completion of CJK joint study on a tripartite FTA, the launch of negotiations is just a matter of time, and CJK FTA and TPP are expected to stand on the same starting line. Compared to TPP negotiations, which are expected to be full of challenges and difficulties, CJK FTA negotiations will start in a more mature and favorable condition. It is very likely the CJK FTA will be up and running earlier than TPP if all parties work relentlessly together.

2. Suggestions

Thanks to reform and opening up, China has scored tremendous achievements in its economic development and social progress. Our two great goals in the first half of this century are to build a moderately prosperous society of higher standards benefiting more than one billion people by 2020 and turn China into a great modern socialist country that is prosperous, strong, democratic, culturally advanced and harmonious by 2050. Given the size of its population, territory and natural endowments, China will surely become a global power by the middle of this century. To ensure that all countries in the world understand and recognize China's development and status, and support China's development process, we need to have the recognition and support of Asian countries in the first place. To this end, China, a country in East Asia, needs to make regional cooperation in East Asia a success first. A sound regional cooperation in East Asia is conducive to both China's own development and East Asia's development as a whole. It will be in the best interests of East Asia and contribute to common and win-win development of all East Asian nations. Therefore, China should remain focused on East Asia and East Asian regional cooperation.

(1) We must seize the opportunity of East Asian cooperation and create an environment in East Asia favorable to our development. Our experience in the reform and opening up prove that the creation of a peaceful and stable peripheral environment is essential for China's

smooth economic and social development. Since the end of the Cold War, China has leveraged the opportunities of the former Soviet Union's disintegration and the Asian financial crisis to its own advantage, and made major breakthroughs in improving relations with countries in Central Asia and East Asia. It is important that we seize the window of strategic opportunity to our favor, leverage our geopolitical, economic and cultural advantages, increase our strategic investment in East Asia, properly handle our relations with ASEAN, Japan, the ROK, the United States and Russia, consolidate China's initiative in East Asia, promote the process of East Asian cooperation and create a more favorable geopolitical and economic environment in East Asia.

(2) We must remove disturbances of all kinds to promote steady progress of East Asian cooperation on the right track. After the US and Russia joined the EAS this year, how to navigate East Asian cooperation has been a matter of great interest to all parties. We should adopt a proactive and open approach to East Asian cooperation. First, we need to let ASEAN play the leading role in East Asian cooperation and encourage ASEAN to adhere to the principle of non-interference in internal affairs and consultation, and the "ASEAN way" of consensus-building. We may build on the 20th anniversary of dialogue between China and ASEAN to enhance political mutual trust with ASEAN. Second, we need to maintain the 10+3 as the main channel of East Asian cooperation and strive for more progress in 10+3 cooperation on the ground. We will make 10+3 cooperation a model for East Asian cooperation and lay the groundwork for stronger and more wide-ranging East Asian cooperation. Third, we need to maintain the EAS as a strategic forum participated and led by national leaders and adopt a more open and flexible attitude towards the Summit agenda and agenda setting. We may help navigate the EAS in keeping with the trend of the times and build it into a strategic platform for enhancing mutual trust and promoting cooperation.

(3) We need to leverage China's advantages in economic growth to contribute to greater East Asian cooperation. East Asian countries are generally positive about the momentum of China's economic growth, and are more willing to develop cooperation with China. The demand for cooperation with China is on the rise. We should seize the opportunity, and take the initiative to plan and expand cooperation with East Asian countries, hopefully that will help expand China's role as the center of

East Asian economic growth. The implementation of China's 12th Five-Year Plan for economic and social development, which prioritizes the expansion of domestic demand, especially demand for consumption, offers us an opportunity to expand imports from East Asian countries. We should firmly implement the "going global" strategy, increasing investment in East Asian countries and economic assistance to developing countries in the region. We should also work towards greater regional cooperation, closer communication and infrastructure connectivity between East Asian nations. We should also promote two-way flow and cooperation with East Asian countries in tourism, education, and culture, and deepen mutual understanding and friendship between the Chinese people and the peoples of East Asia.

(4) We need to build up our capacity and preparations for East Asian cooperation. We should take an active part in the design and development of regional cooperation in East Asia. Fifteen years have passed since the establishment of the 10+3 cooperation mechanism in 1997. China has been an active participant. East Asian cooperation has become an integral part of China's diplomacy towards its neighbors. Our positive attitude and practical contribution are recognized and commended by ASEAN countries. That said, we still need to improve our work in order to play a greater role in the next stage of East Asian cooperation. Firstly, we should strengthen our strategic planning on regional cooperation in East Asia, setting clear goals and making plans for different stages; secondly, we should strengthen inter-agency coordination, mobilize necessary resources, give full play to China's aggregate strength and actively participate in East Asian collaborative activities in all fields; and thirdly, we should strengthen research on issues relating to regional cooperation in East Asia, so as to gain the initiative in the design and setting of the EAS and East Asian cooperation agenda and be in a better position throughout the process of East Asian cooperation.

Promoting Regional Integration Serves the Shared Interests of East Asian Countries

Remarks at the closing ceremony of the Ninth East Asia Forum
Chengdu, September 16, 2011

Enhancing the internal growth momentum in East Asia and promoting regional integration meet the actual needs and common interests of all East Asian countries. Regional countries should speed up the development of an East Asian free trade area and promote intra-regional trade so as to eventually foster a greater regional market. Advancing financial integration is imperative for such a market. To this end, we should continue to enhance the effectiveness of the Chiang Mai Initiative Multilateralization (CMIM) and accelerate the development of the Asian Bond Market to improve the region's investment and financing environment and increase its resilience to financial risks. China, Japan and the ROK should support the implementation of ASEAN's Master Plan on Connectivity. Parties need to work together to deepen cooperation, promote connectivity and create new growth areas. East Asian countries are on different rungs of the industrial ladder. We should make the most of this feature and ensure the success of industrial relocation and transfer to bridge development gaps, optimize resource allocation, realize sound industrial development in the region and advance the process of regional integration.

The Asian financial crisis in the late 20th century engendered 10+3 cooperation. In the face of the sudden outbreak of the crisis, we worked in unity and coordination to accelerate the development of East Asian cooperation frameworks, thus turning a crisis into opportunities and opening up new horizons for common development. Thanks to years of cooperation, we now enjoy a more solid economic foundation, wide-ranging cooperation and more mature regional mechanisms. East Asia now stands as the most dynamic and promising region in the world, an important engine for world economic growth, and an important force for stability and development.

Today, East Asia is undergoing profound transformation. The underlying impact of the international financial crisis and destabilizing factors and uncertainties in the world economy have brought new challenges to East Asia's future development. In today's world, all countries are so interdependent that they rise and fall together and no country can succeed on its own. Countries in East Asia are even more closely linked. Confronted with the complex situation, we should all strengthen solidarity and cooperation, broaden our thinking and tap our potential for common development.

Next year will mark the 15th anniversary of 10+3 cooperation. We should seize the opportunity to review the past and plan for the future. In this regard, the East Asia Forum can provide important intellectual support. The just-adopted Concept Paper of the East Asia Forum has set new targets, put forth new requirements and injected new impetus for the forum. Here, I wish to thank all the countries represented, the ROK in particular, for their efforts to strengthen the forum. I am convinced that the East Asia Forum will continue to contribute ideas and add new dynamism to 10+3 cooperation, and take forward the wider East Asian cooperation.

China has scored remarkable achievements in its development. However, it still suffers from a lack of balance, coordination and sustainability in development. China has formulated and is implementing the 12th Five-Year Plan, which focuses on promoting scientific development, accelerating the transformation of the growth pattern, maintaining long-term, steady and fairly fast economic growth, improving people's lives and enhancing social harmony and stability. China will give priority to expanding domestic demand. It will also open wider to the outside world and continue to follow the win-win strategy of opening-up. China will continue to firmly support ASEAN's integration and community-building efforts, firmly support ASEAN's centrality in regional cooperation and firmly support the sustained development of East Asian cooperation with 10+3 cooperation as the main vehicle. China is ready to work in unison with ASEAN, Japan and the ROK to promote liberalization and facilitation of intra-regional trade and investment, uplift the level of financial cooperation and enhance connectivity to achieve common prosperity.

Standing Together to Cope with Challenges

Assistant Foreign Minister Liu Zhenmin talks about Asian Situation and China's Diplomatic Work with Neighboring Countries
January 8, 2012

On January 8, 2012, the author received an exclusive interview of Xinhua News Agency, during which he introduced the Asian situation and China's diplomatic work with neighboring countries in 2011.

Q: Regional situation in Asia experienced profound changes over the past year. What do you think of the current Asian situation?

A: The world today is in the era of major changes, major readjustments and major development. The economic and political situation in Asia is also undergoing profound changes. In general, stability and development are the ultimate aspiration of countries in the region and cooperation is the mainstream of their foreign policy.

Asia is playing an increasingly important role in the international landscape and has become the most dynamic region with the strongest development potential in the world. First, as the international financial crisis and European debt crisis continue spreading and the world economic and financial situation remains volatile, Asia maintains relatively fast growth. Its economic growth rate is expected to reach 7-7.5 percent in 2011, making it an important engine of world economic development. Second, Asia as a whole remains stable. Although the region is faced with problems, related parties have managed to properly handle and control their differences through friendly consultations and the overall stability in the region remains intact. Third, all Asian countries emphasize mutually beneficial cooperation and desire for common development through cooperation.

On the other hand, Asia is also confronted with many uncertain and unstable factors. Regional hotspot and tough issues are still prominent, economic risk factors are lingering and non-traditional security challenges, such as natural disasters and environmental degradation arc on the rise. We firmly believe that Asia will seize opportunities to make greater contributions to world peace and development so long as we carry forward the Asian spirit of coexistence in harmony, good-neighborly friendship, consultations and dialogue, respect for civilization diversity, solidarity and cooperation and stick to the development roads suitable for our own national conditions.

Q: Currently much of the world opinion is focused on China's relations with its neighboring countries. What do you think of the development of China's relations with neighboring countries?

A: In 2011, China's relations with Asian neighboring countries in general maintained the sound momentum of stable development. Good-neighborly friendship, political mutual trust and mutually beneficial cooperation were further deepened. China and its Asian neighbors overcame the negative impact of the international financial crisis, properly handled regional hotspot issues, actively pushed forward regional cooperation and expanded exchanges and cooperation in the fields of trade, investment, infrastructure, finance, science and technology, culture and education to unprecedented width and depth.

We maintained frequent high-level contacts with other Asian countries. In 2011, we exchanged high-level visits with almost all the Asian countries and Chinese leaders attended a number of important regional multilateral conferences. Chinese President Hu Jintao attended Boao Forum for Asia Annual Conference 2011. Chairman of the Standing Committee of the National People's Congress Wu Bangguo visited Maldives. Premier of State Council Wen Jiabao attended the China-Japan-ROK Tripartite Summit in Japan, visited Indonesia and Malaysia, participated in the meetings of East Asian leaders in Indonesia and visited Brunei. Chairman of the National Committee of the Chinese People's Political Consultative Conference Jia Qinglin visited Myanmar. Vice President Xi Jinping visited Vietnam and Thailand. Vice Premier Li Keqiang visited the DPRK and the ROK. Member of the Standing Committee of the Political Bureau of the Communist Party of China Central Committee Zhou Yongkang visited Nepal, Laos, Cambodia and Mongolia. State Councilor Meng Jianzhu visited Mongolia and Pakistan. State Councilor Dai Bingguo visited Pakistan as the envoy of President Hu Jintao. There were altogether nearly 90 exchange activities above the foreign minister level. Through those activities, we strengthened communications and enhanced understanding and mutual trust with relevant countries and put forward major measures to promote the bilateral pragmatic cooperation, which have effectively boosted the growth of bilateral relations.

China upholds that countries, small or big, are all equals, Asian countries should be respected and supported to choose their own development roads, and disputes and differences should be solved through dialogue and consultations. We have reinforced good-neighborly friendship with Asian neighboring countries. We have raised our relations with Myanmar and Mongolia to the level of "China-Myanmar all-round strategic partnership of cooperation" and "China-Mongolia strategic partnership" respectively. The year 2011 marked the 20th anniversary of China's dialogue relationship with ASEAN and the two sides jointly held celebrations throughout the year. Premier Wen attended the 8th China-ASEAN Expo in Nanning and put forward a series of initiatives and measures on deepening the bilateral economic and trade cooperation, connectivity, social and cultural cooperation at the commemorative summit on the 20th anniversary of China-ASEAN dialogue relationship in Bali, Indonesia. The two sides issued the Joint Statement to plan the future development of bilateral relations. Dai Bingguo participated in the fourth Summit of the Greater Mekong Subregion (GMS)-Economic Cooperation Program in Myanmar, which further deepened China's mutually beneficial and pragmatic cooperation with GMS countries. The year 2011 also marks the 20th, 50th and 60th anniversary of China's diplomatic relations with Brunei, Laos and Pakistan respectively and the "Year of China-India Exchanges". Various celebrations were held between relevant parties, which helped increase political mutual trust and friendship between the people.

In general, China's relations with Asian countries have withstood the test of fluctuating international situations and continue moving forward. Asian countries at large take the development of relations with China as one of their most important foreign relations. Good-neighborly friendship remains the mainstream of Asian countries' policy with China and to step up cooperation with China is the major trend.

Q: In 2011, the economic recovery of developed countries was weak, the European debt crisis sprawled and unstable and uncertain factors of the world economic situation rose significantly. Against such a background, what role has China played to boost the economic growth and mutually beneficial cooperation among Asian countries?

A: At present, the international financial crisis is still exerting in-depth impact and the prospect of global economic and financial situation is not optimistic. In the age of deep economic globalization, no single region or country can seek development alone. We must strengthen cooperation and seek development together. China believes in solidarity and cooperation, mutual benefit and win-win cooperation. We will work with other Asian countries to face up to challenges and strive for common development.

China maintains a sound momentum of trade with Asian countries. From January to November 2011, China's total trade volume with Asian countries reached US$965.2 billion, up 21 percent over the same period of the previous year, its import from Asian countries increased 18.7 percent on a year-on-year basis to US$530.1 billion and export grew 23.9 percent to US$435.1 billion. Its trade deficit with Asian countries was about US$95 billion. China maintained its position as the biggest export market of Asian countries and remained the largest trade partner of the DPRK, Mongolia, Japan, the ROK, Vietnam, Malaysia and India. China's trade structure with Asian countries has been optimized and completed the switch from primary products to manufactured goods. In particular, the percentage of trade in new and high-tech products is rising year by year.

China's investment in Asian countries has grown rapidly. As of November 2011, China's non-financial direct investment in Asian countries totaled US$18.03 billion. Asia became the most concentrated area of "going abroad" Chinese enterprises. China is the no.1 source of foreign investment in Myanmar, Cambodia, the DPRK and Mongolia. It will also build a Free Trade Area (FTA) in each of the ASEAN member countries and lift up economic and trade cooperation through cluster investment.

China's FTA with Asian countries moves forward steadily. China-ASEAN FTA operates soundly since it was comprehensively launched in 2010. The mobility of capital, resources, technologies and human resources between the two sides was improved significantly and the bilateral economic and trade exchanges grew more active. From January to November 2011, trade volume between China and ASEAN reached US$328.9 billion, a year-on-year increase of 25.1 percent, making ASEAN China's third largest trade partner. The government-industry-academia joint research under China-Japan-ROK FTA was completed as scheduled. China and Japan put forward the joint initiative to speed up the development of East Asia FTA and East Asia all-round economic partnership. China also actively pushed forward the FTA negotiations with the ROK and India.

China has deepened fiscal and financial cooperation with Asian countries. In 2011, China signed the bilateral currency swap agreement with Thailand, Pakistan and Mongolia worth RMB70, RMB10, and RMB5 billion respectively and expanded the currency swap with the ROK to RMB360 billion. The total amount of bilateral currency swap agreements China signed with Asian countries reached RMB775 billion. The Industrial and Commercial Bank of China and China UnionPay set up branches in Laos, Singapore, Pakistan and India for business expansion.

China cooperated with Asian countries closely in the sectors of new and high-tech, new energy, environmental protection and energy saving. China-Japan Caofeidian Ecological Industrial Park and Lianyungang Ecological Science and Technology Industrial Park were kicked off and China-Singapore Eco City program went on well. A number of cooperation agreements were reached at the sixth China-Japan Energy Efficiency and Environmental Protection Forum and the first China-Japan Green Expo. China, Japan and the ROK strengthened cooperation on renewable energy and energy efficiency. China officially launched the construction of "two economic zones" with the DPRK and advanced the cooperation programs of Knowledge City, food zone and economic and trade zone with Singapore, Vietnam and Cambodia.

China stepped up assistance to Asian countries. In March 2011, Japan suffered the triple disasters of earthquake, tsunami and nuclear leakage. China provided Japan with RMB30 million of emergency assistance materials and 20,000 tons of fuel oil. When Thailand and Cambodia suffered the most serious flood, the Chinese government offered timely aid of materials and remittance and dispatched expert team to Thailand for disaster relief consultation. When disasters took place in the DPRK, the ROK, the Philippines, Pakistan and Sri Lanka, the Chinese government and Red Cross also provided timely humanitarian assistance. China continued providing the less developed countries, including

Nepal, Sri Lanka, Maldives, Laos, Bangladesh and Afghanistan with aid or technical training to help them improve self-development capacity.

In addition, China is committed to promoting the connectivity with Asian countries in the fields of road, railway, telecom and port. On the basis of promising US$15 billion of credit in 2009, China added another US$10 billion of credit to support ASEAN's infrastructure connectivity and bring benefit to all ASEAN member countries.

Q: Japan and India are important neighbors of China. Japanese Prime Minister visited China lately and China and India are preparing another round of meeting of their special representatives of border issue. What does the Chinese side think of relations with Japan and India?

A: The year 2011 witnessed the improvement and stable development of China-Japan relations. After the March 11 massive earthquake broke out in Japan, the Chinese government, leaders and people extended condolence and support to the Japanese people through various ways. President Hu personally went to the Japanese embassy in China to express grief over the victims, which is unprecedented in the history of China-Japan relations. When Premier Wen attended the fourth China-Japan-ROK Tripartite Summit in Japan, he especially went to Miyagi and Fukushima to visit the disaster-hit people and reached a series of important consensus with Japanese leaders on strengthening post-disaster reconstruction cooperation. Those measures fully demonstrated the goodwill of the Chinese government and people.

After the new Japanese cabinet was formed, President Hu and Premier Wen held meeting and talks with Prime Minister Yoshihiko Noda during the G20 Summit in Cannes, the APEC Economic Leaders' Meeting in Hawaii and the East Asian leaders' meetings in Bali, setting a good start for the bilateral relations. Not long ago, Prime Minister Yoshihiko Noda visited China successfully. Both China and Japan attached great importance to the visit. President Hu and Chairman Wu Bangguo met with him respectively and Premier Wen held talks with him. The two sides reached broad consensus in the political, economic and cultural sectors. The visit achieved full success and boosted the progress of bilateral strategic and reciprocal relations.

China-Japan exchanges and cooperation have achieved major

progress in various fields. The bilateral trade volume exceeded US$300 billion last year, hitting another historic high and making China Japan's largest trade partner for the third consecutive year. People of the two countries overcame the difficulties caused by the triple disasters and people-to-people contacts maintained the high level of over five million people. More than 247 pairs of cities between the two countries established sister city relations.

2012 marks the 40th anniversary of the normalization of China-Japan diplomatic relations and the "Year of Friendly Exchanges between the Chinese and Japanese People". China expects to take this opportunity and join hands with Japan to enhance political mutual trust, appropriately handle differences and problems, step up dialogue, exchanges and cooperation in various fields and at all levels and promote the healthy and stable progress of bilateral strategic and reciprocal relations according to the principles of the four political documents and the important consensus reached between leaders of both countries.

China-India strategic partnership of cooperation has maintained the momentum of healthy and stable development in recent years. The two sides keep frequent high-level contacts. The bilateral exchanges and cooperation are strengthened. China and India enjoy smooth communications and coordination on major regional and international issues. The China-India relationship featured by sustained, stable and healthy development not only benefits the two peoples but also contributes to peace, stability and prosperity in Asia and the world at large.

2011 was the "Year of China-India Exchanges". The two countries held colorful exchange activities which enhanced mutual understanding and deepened friendly cooperation. President Hu and Premier Wen held bilateral meetings with Prime Minister Singh on international occasions. During the first 11 months, China-India trade volume exceeded that of the whole year 2010 and grew by 21.8 percent to US$67.28 billion. China and India held the first strategic economic dialogue, fiscal and financial dialogue and defense and security consultations and 500 young Indian delegates visited China. Those major exchange activities and pragmatic cooperation effectively strengthened the bilateral relations.

In 2012, China expects to make joint efforts with India to continue to implement the consensus reached between leaders of both countries, maintain high-level exchanges, step up strategic mutual trust, deepen mutually beneficial cooperation in various fields, properly handle problems of the bilateral relations, support and learn from each other and promote the development of China-India strategic partnership of cooperation.

Q: Some regional hotspot issues, like the South China Sea and the Korean Peninsula nuclear issues became fairly prominent recently. What has China done to maintain regional peace and stability as a major country in the region?

A: As a responsible major country, China advocates the new security concept of mutual trust, mutual benefit, equality and collaboration, takes an active part in seeking solution of regional hotspot issues and is committed to working with Asian countries to build a region of peace and stability, equality and mutual trust and win-win cooperation.

At present, the situation in the South China Sea is peaceful and stable in general. With regard to certain disputes over some islands and reefs and maritime delimitation in the South China Sea, China always upholds that the disputes shall be settled peacefully by parties directly concerned through negotiations. It is also the consensus reached among relevant countries. Parties can put aside the disputes before solutions are found and make common development. External forces shall not intervene in the South China Sea disputes with any excuse.

In 2011, China and ASEAN member countries reached agreement on the guiding principles of implementing the Declaration on the Conduct of Parties in the South China Sea, launched pragmatic cooperation within the framework of the Declaration and established the China-ASEAN Maritime Cooperation Fund. China and Vietnam signed the Agreement on the Basic Principles Guiding the Resolution of China-Vietnam Maritime Issues. China also formed the maritime cooperation committee and maritime cooperation fund with Indonesia. It fully proves that China and ASEAN member countries have the determination, wisdom and capability to jointly maintain peace and stability in the South China Sea.

South China Sea is an important international transport channel. Navigation in the South China Sea is free and sea-routes there are safe. They have never been affected by the dispute. China attaches great importance to navigation freedom and safety in South China Sea. It actively participates in and is committed to the safety of navigation in the South China Sea. China and ASEAN member countries have held the seminar on South China Sea navigation freedom and safety and parties have reached extensive consensus over the issue.

China always pays close attention to the situation on the Korean Peninsula and consistently upholds that the Korean Peninsula nuclear issue be solved through dialogue and consultations and in peaceful ways. In 2011, related parties increased interactions on the resumption of Six-Party Talks and made efforts to ease the situation on the Peninsula. China always believes the Six-Party Talks are an efficient mechanism of realizing denuclearization on the Peninsula and maintaining peace and stability on the Peninsula and in Northeast Asia as well as an important platform for related parties to improve relations. To retain the Six-Party Talks complies with the interest of all parties concerned. I hope the parties concerned build confidence and continue to keep dialogue and stay in touch. Holding presidency of the Six-Party Talks, China will continue to promote peace talks and work with all the related parties to play a constructive role of advancing the Six-Party Talks process and build lasting peace and tranquility on the Peninsula and in Northeast Asia.

Q: The DPRK's top leader Kim Jong Il passed away not long ago and the situation of Northeast Asia has drawn wide attention. What role will China play under the current circumstances? What's the expectation of China for the development of relations with the DPRK?

A: It is in alignment with the interest of all parties and the common expectation of the international community that the DPRK maintains stable development, the Korean Peninsula remains peaceful and stable and Northeast Asia realizes lasting peace and tranquility.

China and the DPRK are friendly neighbors linked by mountains and rivers. The DPRK is an important party to the issues of the Peninsula and Northeast Asia. After Chairman Kim Jong Il passed away, the Chinese side expressed grief and condolence through many ways. We believe that the DPRK people will turn sorrow into force and continue to push forward the DPRK's socialist undertaking under the leadership of the Workers' Party of Korea and Comrade Kim Jong-un.

We have actively communicated with all the related parties including the DPRK and clearly elaborated China's position and concerns of supporting the DPRK's stable development and firmly maintaining peace and stability of the Peninsula and Northeast Asia. Our communication with related parties is efficient and we have reached wide consensus on maintaining regional peace and stability.

It is the consistent principle of the Chinese Communist Party and the Chinese government to consolidate and develop friendly and cooperative relations with the DPRK. We will carry forward the spirit of "tradition inheritance, future orientation, good-neighborly friendship and strengthening cooperation" to pass on and develop China-DPRK relations. We will, as always, provide support and help within our capacity for our friendly neighbor – the DPRK.

Q: The Sixth Plenary Session of the 17th CPC Central Committee put forward the target of building socialist cultural power with Chinese characteristics. What has China done to promote social and cultural exchanges with Asian countries? What achievements have been made?

A: China and Asian countries are close to each other geologically and interlinked culturally and enjoy unique advantages and huge potential of conducting social and cultural exchanges. The Sixth Plenary Session of the 17th CPC Central Committee made the strategic decision of raising the level of cultural opening-up and promoting the Chinese culture to the world and pointed out the direction for China to step up social and cultural cooperation with its neighboring countries. The year 2011 witnessed the in-depth development of China's cultural exchanges with Asian countries and the deepening of mutual understanding and friendship between the Chinese people and people of other Asian countries.

We held a number of celebrations with related countries on the 20th anniversary of China-Brunei diplomatic relations, the 50th anniversary of China-Laos diplomatic relations, the 60th anniversary of China-Pakistan diplomatic relations, China-Pakistan Friendship Year and the Year of China-India Exchanges, which effectively reinforced the friendship between the people. In order to commemorate the 20th anniversary of China-ASEAN dialogue partnership, the two sides held more than 40 celebration activities, including the exchange of congratulation messages between the leaders, exchange of visits of the media, cultural exchanges and exchanges of youth, receiving warm responses from ASEAN member countries and making the China-ASEAN friendship from generation to generation deeply rooted in the heart of the people.

We organized the China-Japan film week and animation festival, sent Buddhist tooth to Myanmar for local people to worship, held Asian cultural cooperation forum and Asian cultural festival and received a large number of Japanese and Indian youth, which have all consolidated the friendship between the Chinese people and people all over Asia.

Personnel contacts between China and Asia are also on the rise. From January to September 2011, the personnel contacts between China and ASEAN member countries reached 10.09 million people. 10.46 million people from 13 Asian countries among which are Mongolia, the ROK, Japan, Singapore and Malaysia visited China, making up over 50 percent of the inbound foreigners in China. China and ASEAN are committed to increasing tourism cooperation and will work hard to realize the objective of 15 million people of personnel contacts between the two sides by 2015.

China has signed cultural agreement or cooperation memorandum with most Asian countries. More than one third of China's external cultural exchange programs are implemented with Asian countries. China has signed educational cooperation agreement or MOU with nearly all Asian countries. Bilateral and multilateral education exchanges and cooperation within the region are well underway. Education and training are expanding rapidly. Forty percent of the scholarships of Chinese government are granted to Asian countries and the quota is still rising. China has established 83 Confucius Institutes and 40 Confucius Classrooms in more than 30 Asian countries, dispatched over 1,400 Chinese teachers and nearly 10,000 volunteers and trained tens of thousands of Chinese language teachers for Asian countries.

The landing of mainstream Chinese media, like Xinhua News Agency, CCTV and CRI in Asian countries are going on and the number of overseas stations of correspondents, studios and staffs are increasing. CCTV channels have landed in 27 Asian countries, with around 35.9 million of subscribers of the English channel and 98 million subscribers of the Chinese channel. CRI has landed in 11 Asian countries. Border provinces such as Guangxi, Yunnan and Inner Mongolia use their geological and cultural advantages to conduct media cooperation with the neighboring countries and launched Tibetan, Kazak, Uygur and Mongolian language broadcasting in some neighboring countries. Cultural programs such as Chinese Language Bridge, Feeling China and Happy Chinese New Year are widely welcomed. China's power of cultural communication in Asia keeps rising.

Q: The joining of the US and Russia in East Asia Summit this year marks a new stage of East Asia regional cooperation. How does China view and what does it think of the current East Asia regional cooperation process? What role has China played to promote East Asia regional cooperation?

A: China actively participates in and spares no efforts to promote East Asia cooperation and has made important contributions to the development of regional cooperation. China took the lead among major countries in joining the Treaty of Amity and Cooperation in Southeast Asia and building the China-ASEAN FTA covering a total population of 1.9 billion. China vigorously pushes for the Chiang Mai Initiative Multilateralization, ASEAN and China, Japan and the ROK rice emergency reserve and the East Asia FTA and the cooperation has achieved fruitful results in the fields of finance, economy, trade and food security with relevant parties. China also pushed forward East Asia Summit to discuss the major strategic topics concerning East Asia's peace and development and provided strategic guidance in the discussion.

East Asia cooperation is developing soundly, various cooperation mechanisms move forward in parallel, pragmatic cooperation has made new progress and tangible benefits have been brought to East Asian people. In November 2011, East Asian leaders' meetings were held in Bali, Indonesia successfully. The meetings upheld the theme of solidarity, development and cooperation, the 10+1 and 10+3 cooperative mechanisms and adhered to seeking solution of problems between countries through peaceful ways and consultations. China is full of confidence in the future of East Asia cooperation.

Currently, the in-depth impact of international financial crisis is still spreading and the economic downside risks of countries in the region are on the rise. The historical issues of East Asian countries are not solved yet and non-traditional security threats are increasing. East Asian countries are highly diversified and the regional cooperation architecture is going through complicated changes. Under such circumstances, China believes that East Asia cooperation should maintain openness and inclusiveness and at the same time adhere to the established direction of various mechanisms, insist on the effective principles and models and stick to the leading role of ASEAN. East Asian countries should be fully motivated in the process. The East Asian cooperation architecture with 10+1 as the basis, 10+3 as the main framework and the East Asia Summit as the key supplement should be carried forward. The theme of development, cooperation, mutual benefit and win-win outcome should remain unchanged. Relevant cooperation should be pushed forward step by step in the direction of building regional peace, stability and prosperity.

Q: Asia-Pacific security cooperation this year drew close attention from various countries. As an important member of ASEAN Regional Forum and other regional security cooperation mechanisms, what kind of regional security cooperation architecture does China think should be built in the region? What role will China play in this architecture?

A: The regional security architecture is constantly developing and evolving and reflects the real needs of regional security. After years of explorations and efforts, Asia-Pacific has formed the multi-level and compound security cooperation architecture. The Six-Party Talks, the Shanghai Cooperation Organization (SCO), the ASEAN Regional Forum and the ASEAN Defense Ministers Plus Forum are all progressing side by side. In the short run, it is difficult to form a pan-Asia-Pacific security cooperation mechanism overriding all the other mechanisms in the region. Existing mechanisms should continue to give full play to their respective advantages, complement and promote each other and play a constructive part.

The regional security architecture should be based on the new security concept of mutual trust, mutual benefit, equality and collaboration. Asia-Pacific is a highly diversified and complicated region. To conduct security cooperation in the region, the first issue to be addressed is the lack of trust between related countries. States should respect each other, make equal consultations, understand and accommodate each other, take care of each other's interest and concerns and seek together mutually beneficial cooperation and development. Countries should seek harmonious coexistence while maintaining diversity, seek common ground while putting aside differences, expand consensus step by step, eliminate differences, push forward cooperation and jointly safeguard regional peace, stability and development.

The regional security architecture should be open, transparent, inclusive and representative. Non-traditional security issues, such as terrorism, climate change, food security, energy security and maritime security are growing increasingly prominent in Asia-Pacific and hotspot issues like the Korean Peninsula nuclear issue keep popping up. Security issues become more abrupt, conductive and correlated. The closed and exclusive security cooperation arrangements will be more and more unable to cope with such global and compound challenges. Under the new circumstances, Asia-Pacific countries must abandon the cold-war mentality, conduct equal dialogue, consultations and cooperation to truly solve the problems and differences. Seeking peace through cooperation, ensuring security through cooperation, solving conflicts through cooperation and promoting harmony through cooperation should be the basic principle countries follow when addressing security issues.

Over the years, China has been committed to maintaining peace and stability, conducting mutually beneficial cooperation and building common prosperity in Asia-Pacific. China raised the initiatives of building the Six-Party Talks and the SCO and played a critical role for peace, stability and development of the region. China is a key member of and actively promotes the ASEAN Regional Forum, ASEAN Defense Ministers Plus Forum, Council for Security Cooperation in the Asia-Pacific and the Shangri-La Dialogue, raised a series of cooperation initiatives, hosted multiple cooperation programs and received appreciation from the relevant parties. China is willing to strengthen cooperation with other countries in the future to raise mutual trust, step up cooperation, maintain regional peace and stability and build an open, transparent and equal Asia-Pacific security architecture.

Q: China's diplomacy with neighboring countries achieved major progress in 2011. Could you please describe China's diplomatic work with Asian countries in 2012?

A: The situation in the region will continue to evolve in 2012. China's diplomatic work with Asian countries will be more onerous. We will adhere to the principle of being a good neighbor and a good partner of neighboring countries, increase political mutual trust, expand common interest, strengthen exchanges and cooperation and handle differences and problems properly with Asian countries and work with them to forge a regional environment with peace and stability, equality and mutual trust and win-win cooperation.

We will further deepen good-neighborly friendship with Asian countries. The nuclear security summit in the ROK, the BRICS summit in India, the Asia Europe Meeting in Laos and the East Asian leaders' meetings in Cambodia are all important occasions for the Chinese leaders to meet with the leaders of other Asian countries. We will make full use of the exchanges of high-level visits, strategic dialogues, consultations and other mechanisms with relevant countries to enhance strategic communications and mutual understanding. We will continue to play a constructive role in solving regional hotspot issues, appropriately address and settle differences and problems with neighboring countries through dialogue and consultations and maintain regional peace and stability jointly.

We will deepen the convergence of interest with Asian countries. We will continue to implement the mutually beneficial and win-win strategy of opening-up, accelerate the connectivity of infrastructure, boost regional fiscal and financial cooperation, encourage competitive Chinese enterprises to invest in neighboring countries, advance the FTA with related countries and overseas economic cooperation zones and raise the level of economic and trade cooperation with surrounding countries.

We will further strengthen cultural exchanges with Asian countries. In 2012, we will make elaborate preparation for the celebrations on the 40th anniversary of the normalization of China-Japan diplomatic relations, the 40th anniversary of China-Maldives diplomatic relations, the 20th anniversary of China-ROK diplomatic relations and the tenth anniversary of China-East Timor diplomatic ties, actively carry out public diplomacy and enhance mutual understanding and friendship between the people. We will continue to build Chinese Cultural Center and Confucius Institutes and Classrooms in Asian countries and broaden cultural exchanges with them.

China's diplomatic work with neighboring countries enjoys bright prospects in 2012. We will face difficulties, keep forging ahead, make innovations, explore fearlessly and fully devote ourselves to the diplomatic work with neighboring countries so as to build a sound environment for domestic economic and social development and make active contributions to promoting peace and prosperity in Asia and the whole world.

East Asian Cooperation: Challenges and Opportunities

Speech at the Geneva Center for Security Policy
October 24, 2012

Today I wish to talk about East Asian Cooperation: Challenges and Opportunities. Why have I chosen this topic? First of all, I come from East Asia. Secondly, it is because the current situation in East Asia has drawn wide attention and diverse interpretations.

Some are of the view that, against the backdrop of the global financial crisis, East Asia has retained relatively stable economic development and become a bright spot in the crisis as well as an important engine of global growth. Regional cooperation in East Asia continues to deepen and promises a bright future. Some others believe that, given the shadow cast by the recent geopolitical tension, historical feud and ongoing grievances, East Asian regional cooperation offers little room for optimism. So, how do we assess the current situation in East Asia and the prospect of East Asian cooperation? I wish to share with you my observations from the following four aspects.

I History and current status of regional cooperation in East Asia

During the Cold War, East Asian countries were split into two camps in the polarized world, making regional cooperation too much of a luxury. The Association of Southeast Asian Nations (ASEAN) was initiated with the focus on political and security issues.

The end of the Cold War removed the biggest political obstacle and became a watershed for regional cooperation in East Asia. As it took in more members, ASEAN launched separate dialogues with China, Japan and the Republic of Korea (the ROK), known as the "10+1" dialogues. In 1992, ASEAN decided to build a free trade area, shifting its focus to economic cooperation in order to improve the competitiveness of its member states in the global economy to embrace the tide of globalization.

The Asian financial crisis in 1997 became a catalyst in speeding up regional cooperation in East Asia. Countries in East Asia realized that regional cooperation is not only a must in the sweeping globalization, but also an important way to boost collective strength and promote common development in the region. In 1997, ASEAN launched the 10+3 framework with China, Japan and the ROK, putting East Asian cooperation on a fast track that has led to great achievements in the following ten years and more.

1. Fruitful trade and economic cooperation

I would like to use three sets of figures to illustrate the achievements of East Asian cooperation. First, the share of East Asia in the global economy was only six percent in the early 1990s. Today, it is almost 18 percent. Second, trade among East Asian countries only accounted for around 45.37 percent of the region's total in 1992. The share grew to 52.23 percent in 2007. Third, free trade area was non-existent in East Asia in the early 1990s. But in 2008, the ASEAN Free Trade Area was established. Two years later in 2010, five free trade areas, namely ASEAN-China, ASEAN-Japan, ASEAN-ROK, ASEAN-India, ASEAN-Australia-New Zealand free trade areas were launched, benefiting 3.5 billion people in the region. In 2010, eight of the world's top 45 economies were in East Asia, including five ASEAN countries plus China, Japan and the ROK.

In the fight against the Asian financial crisis of 1997 and the international financial crisis of 2008, East Asian cooperation played an important part and demonstrated strong vitality. China, Japan, the ROK and ASEAN launched the Chiang Mai Initiative (CMI), which was later multilateralized, and now boasts a regional foreign exchange reserve pool of US$120 billion and a US$700 million regional Credit Guarantee and Investment Facility (CGIF). The initiative has played an important part in maintaining economic and financial stability in East Asia.

2. Growing security cooperation

With deepening regional economic cooperation and converging

interests of all parties, security dialogue and cooperation among East Asian countries have also gathered pace. First, peace and stability has been maintained in East Asia. The region has seen no hot war since the Korean War and the Vietnam War in the Cold War era. Second, regional hotspots have been put under control. The Six-Party Talks on the Korean Peninsula nuclear issue is now an important platform to ease tension and promote denuclearization on the Peninsula. Countries in the region have in general chosen diplomatic and peaceful means to resolve disputes. Third, non-traditional security cooperation has been effective. From counter-terrorism cooperation to jointly fighting pirates in the Malacca Strait, from disaster relief efforts in the wake of the Indian Ocean Tsunami to the helping hands extended to China after the Wenchuan Earthquake and to Japan after the earthquake and tsunami, the scope and depth of cooperation has grown continuously. Last December, China, Laos, Myanmar and Thailand launched joint patrols along the Mekong River.

3. Flourishing institutional build-up

When the Cold War ended, there was only one regional cooperation framework in East Asia, i.e., ASEAN. After more than two decades, East Asian cooperation is now structured in concentric circles with their respective focus and ASEAN at the center.

10+3 for practical cooperation. Since the inception of the 10+3 cooperation framework in late 1997, more than 50 dialogue frameworks have been established under the umbrella, covering more than 20 areas including finance, economy, trade, science and technology, connectivity and culture. The three 10+1 frameworks with China, Japan and the ROK also complement the 10+3 framework.

10+X for strategic dialogue. The East Asia Summit (EAS) was launched in 2005. Besides the 10+3 countries, the EAS membership also included Australia, New Zealand and India, with the United States and Russia joining in 2010. It is often referred to by the academia as the 10+X. Different from the 10+3, the EAS is a forum spearheaded by leaders of the member states, engaging in strategic dialogue on regional political and economic issues.

ASEAN Regional Forum (ARF) for enhancing mutual trust. The ARF was founded in 1994 and has become a platform for relevant parties to discuss regional hotspots, build confidence, carry out preventive diplomacy and explore ways to resolve conflicts. The forum is broadly based, with 27 members including the 10+X countries, as well as Pakistan, Bangladesh, Sri Lanka, Timor-Leste, the DPRK, Mongolia, Papua New Guinea, Canada and the European Union.

Sub-regional & trans-regional cooperation advancing together. Since 1999, the leaders of China, Japan and the ROK have been holding trilateral meetings regularly, supported by meeting mechanisms at the ministerial, senior officials and working levels. The three countries decided to launch FTA talks in 2012. The Greater Mekong Sub-regional Cooperation (GMS) initiated by the Asian Development Bank (ADB) was launched in 1992, covering six countries, namely China, Cambodia, Laos, Myanmar, Thailand and Vietnam. The GMS practical cooperation now covers a wide range of areas including transportation, energy, telecommunications, environmental protection, tourism, human resource development, trade and investment. The Greater Tumen Initiative was launched by the United Nations Development Program (UNDP) in 1991, with the participation of China, the DPRK, the ROK, Russia and Mongolia. It provides a platform for Northeast Asian sub-regional economic cooperation. In 1996, the Asia-Europe Meeting (ASEM) was launched by ASEAN+3 and the EU and has become a platform of dialogue and cooperation between Asia and Europe. APEC, which was launched in the early 1990s, also covers the majority of countries in East Asia.

II Role of China in East Asian cooperation as an active advocate and contributor

In 1991, China established dialogue relations with ASEAN. The country is both the earliest to champion East Asian cooperation, and an active player and beneficiary of the process.

In the past two decades, thanks to joint efforts, China and ASEAN have gone through an extraordinary path, from overcoming mistrust to engaging in dialogues and building mutual confidence, and then to establishing a strategic partnership for peace and prosperity.

Many "firsts" tell China's role in promoting East Asian cooperation. Among big countries, China was the first to join the Treaty of Amity and Cooperation in Southeast Asia, the first to establish a strategic partnership with ASEAN, the first to explicitly support the Southeast Asia Nuclear Weapon-Free Zone Treaty, and the first to initiate an FTA with ASEAN, which covers a population of 1.9 billion.

China's economic and trade cooperation with ASEAN continues to deepen. In 1991, two-way trade between China and ASEAN was only US$7 billion. Last year, it reached US$362.8 billion, registering an average annual growth rate of over 20 percent. During the 1997 Asian financial crisis, by keeping its currency from depreciation, China took concrete action to support ASEAN countries in their efforts to overcome difficulties. Since the China-ASEAN FTA came into effect, China has been ASEAN's largest trading partner for three consecutive years, and ASEAN China's third largest trading partner. Two decades ago, durian and other tropical fruits produced by ASEAN countries were hard to find in China. But now they can be bought in most Chinese supermarkets.

China is an active participant in East Asian security cooperation. In 2002, Chinese and ASEAN leaders signed the Joint Declaration on Cooperation in the Field of Non-Traditional Security Issues, starting comprehensive cooperation in the area of non-traditional security. Such cooperation has expanded into diverse areas including defense, disaster relief, prevention and control of infectious diseases, counter-terrorism, peacekeeping, protection of sea lanes, and fight against transnational crimes. China has also actively participated in the ASEAN Regional Forum (ARF), the Shangri-La Dialogue and other bilateral and multilateral security dialogues, defense consultations, and joint military exercises and training. China signed with ASEAN countries the Declaration on the Conduct of Parties in the South China Sea (DOC) and agreed on the follow-up guidelines to implement the Declaration.

China is actively engaged in cultural and people-to-people exchanges in East Asia. Today, the two-way personnel flow between China and ASEAN countries has exceeded 13 million annually, and is expected to reach 15 million in 2015. A total of 100,000 exchange students are expected from each side by 2020. In recent years, China has trained for ASEAN countries more than 10,000 professionals in various fields such as trade, telecommunications, agriculture, finance, and water conservancy.

China actively promotes trilateral cooperation among China, Japan and the ROK. China has been ROK's biggest trading partner for years and Japan's biggest trading partner for the past three years.

III The challenges facing East Asian cooperation

Ladies and gentlemen, looking back, we can see that despite a late start, East Asian cooperation has achieved rapid development. At the same time, we should not lose sight of the uneven path it has taken and the many old problems and new challenges it faces.

Number one is regional divergence in East Asia. Being a vast region, East Asia is home to countries at different development stages. We have developed countries like Japan, rapidly growing developing countries as well as the least developed countries. In terms of political system, we have socialist countries, capitalist countries and monarchies. In terms of religion and culture, this region accommodates Taoism, Buddhism, Islam, Catholicism, Christianity and Shinto. Unlike Europe and other parts of the world, the great divergence of East Asia makes regional integration highly demanding.

Number two is the complex and intertwining problems. On the one hand, East Asia is the only region in the world today with the remnant of the Cold War. The situation on the Korean Peninsula constantly strains regional security. Longstanding territorial, ethnic, religious and historical disputes still linger, dampening cooperation from time to time. In recent years, East Asia and the Asia-Pacific region has become the place where major powers interact most intensively and their interests are most intertwined. This has captured the attention of the whole world.

On the other hand, East Asian countries face increasingly salient new common challenges such as terrorism, transnational crimes, natural disasters and infectious diseases.

Number three is institutional building that needs to be further improved. Currently, cooperation frameworks such as 10+1, 10+3 and the East Asia Summit are all informal forums in nature. While boasting the advantage of flexibility, openness and accommodation to the comfort levels of all parties, these frameworks need to be more binding and efficient. The lack of regional security framework is also a weak link for regional cooperation in East Asia.

Number four is the challenge of coordination among the frameworks. The many East Asian cooperation frameworks born in recent years have different focuses. But they overlap in membership and agenda, and in some cases compete with each other, which is not good for pooling resources. In addition, coordination is also needed between East Asian cooperation and Asia-Pacific cooperation.

Number five is how to handle relations with the United States. The United States is an important Pacific country and the only superpower in the world. It is also a member of the EAS and ARF. In theory, East Asian cooperation is a matter for East Asian countries. But the attitude of the United States, positive or negative, affects the process.

IV The Way forward for East Asian Cooperation

Geneva is a center of multilateral diplomacy, and was home to the headquarters of the first universal international organization – the League of Nations. Unfortunately, the League of Nations failed to achieve the shared vision of peace and development for mankind. Built on the ruins of the Second World War, the United Nations has maintained its vitality and become the world's most authoritative and influential international organization. The different destinies of the two organizations, set up in the same century, offer us inspirations on East Asian cooperation. That is, we must adhere to the spirit of peace, win-win cooperation and inclusiveness.

First, East Asian countries should commit to peaceful coexistence, mutual respect, treating each other as equals, and building political mutual trust. The achievements in this region are attributable to the generally peaceful and stable environment in the past decades, which should be cherished by all. In recent years, territorial and maritime disputes among some East Asian countries have been put under the spotlight. They have damaged political trust between parties concerned and affected regional cooperation to some extent. Faced with these problems, countries in the region must never return to the Cold War or zero-sum mentality. They should adopt a vision of common security and shared development; adhere to resolving differences peacefully through dialogue and negotiations, laying a solid political foundation for deepened regional cooperation. This is also what they can draw from the experience of sustained European integration.

In my diplomatic career, I have been involved in border and maritime affairs and know very well how complicated and sensitive territorial issues are as they bear on the sovereignty and security of a state. But

I am convinced that, as long as the parties concerned could bear the bigger picture of cooperation in mind, stick to peaceful negotiations, adhere to mutual understanding and accommodation and always remain patient, the problems confronting them can eventually be resolved properly. China has 14 neighbors on land and eight at sea, with a land boundary of 22,000 kilometers and a continent coastline of over 18,000 kilometers. In 1949 when the People's Republic of China was founded, there was not even one boundary demarcated. In the past 63 years, the Chinese Government has solved boundary issues with 12 land neighbors through peaceful negotiations. The negotiations with India and Bhutan are also underway. On maritime cooperation, China has engaged in dialogues and consultations with all its neighbors. As early as late last century, China signed fishery cooperation agreements with the ROK and Japan respectively. In 2000, China and Vietnam signed the treaty on the delimitation of the Beibu Bay maritime boundary. In 2005, China and the DPRK signed an agreement on the common development at sea. Last October, China and Vietnam signed the Agreement on the Basic Principles Guiding the Resolution of Maritime Issues. These are good examples of China's efforts to solve disputes and promote cooperation through peaceful negotiations.

I know the issues of South China Sea and Diaoyu Islands are very much on your mind recently. I would like to take this opportunity to share with you my views.

The issue of South China Sea is a matter between China and some of its neighbors. Its emergence has also been connected with the new order of the Law of the Sea. On this matter, China has been committed to safeguarding peace and stability as well as freedom of navigation in the South China Sea. On the disputes of sovereignty over some islands and reefs, and differences of maritime boundary delimitation, China has consistently championed proper resolution through negotiations by countries directly concerned. Pending resolution, parties should shelve disputes and engage in common development. In 2002, China and ASEAN countries signed the DOC, committing themselves to resolving the disputes peacefully and safeguarding regional peace. In July 2011, China and ASEAN countries reached consensus over the guidelines of the implementation of the DOC, and started pragmatic cooperation within the DOC framework. At present, the key for all parties concerned in addressing the issue of South China Sea is that they should abide by the spirit of the DOC in real earnest, and refrain from taking actions that may complicate or escalate the disputes and undermine regional peace and stability.

Diaoyu Islands is an issue between China and Japan left over from history. The recent actions taken by the Japanese side have seriously undermined Sino-Japanese relations. The Diaoyu Islands and its affiliated islands have been an inherent part of the Chinese territory in historical, geographic and legal terms. As early as the Ming and Qing dynasties (14th century to early 20th century), Diaoyu Islands was already part of the Chinese territory, and was under Chinese jurisdiction as affiliated islands to Taiwan Island. In 1895, Japan stole these islands by taking advantage of the Sino-Japanese War and forced the Qing Court (the then Chinese government) to sign the unequal Treaty of Shimonoseki and cede to Japan "the island of Formosa (Taiwan), together with all islands appertaining or belonging to the said island of Formosa," where the Diaoyu Islands were included. In 1900, Japan changed the name of the Diaoyu Islands to "Senkaku Islands". In 1941, China declared war on Japan and abrogated all treaties between the two countries. After the Second World War, according to the Cairo Declaration, the Potsdam Proclamation and other international legal instruments, the Diaoyu Islands and other Chinese territories were returned to China. Before 1972, there was no diplomatic relations between China and Japan. In the negotiations of normalization of relations between the two countries in 1972 and the Sino-Japanese Treaty of Peace and Friendship in 1978, leaders of both countries, proceeding from the overall interests of bilateral relations, agreed not to raise the Diaoyu Islands issue for the time being and leave it for solution in the future. In September this year, by taking actions of "purchasing" and "nationalization" of the Diaoyu Islands, the Japanese government has seriously encroached upon China's sovereignty. It is also an outright denial of the outcomes of the victory of the world anti-fascist war and poses a grave challenge to the post-war international order and the purposes and principles of the Charter of the United Nations. The recent moves of the Japanese government to mislead international public opinion can in no way change the historical fact that Japan "stole" the Diaoyu Islands from China in 1895, the fact that these islands were returned to China at the end of the Second World War, and the fact that China has territorial sovereignty over them. The Chinese Government's resolve to safeguard China's sovereignty and territorial integrity is firm. We hope that Japan immediately stops all activities that violate China's territorial sovereignty, take concrete actions to correct its mistakes, and return to the track of negotiation with China for resolving the dispute and promoting common development.

Second, East Asian countries should pursue common development through mutually beneficial cooperation. The past 20 years has witnessed the rapid development of Asia, as well as the take-off of regional cooperation in East Asia. East Asian countries have become a highly interdependent community of interests and destiny, fully leveraging their comparative advantages. Affected by the sluggish global economic recovery, persistent turbulence in the international financial market and rising protectionism, the momentum of growth in East Asia is slowing down. At this crucial stage of development, East Asian countries should take economic development and social progress as their top priority, and adhere to the spirit of mutual benefit and win-win solution. Only by doing so can they effectively mobilize forces, consolidate public support for cooperation and achieve full revitalization of East Asia.

Thirdly, East Asian countries should embrace openness and inclusiveness, making full use of complementary advantages. The diversity of East Asia both poses challenges for and gives strength to cooperation. All parties should have an open mind, respect the lawful presence and legitimate interests of others in the region, embrace constructive contribution from any party, and learn from the useful experience of integration of Europe and other regions. They should accept the divergence and diversity of the region with a broad mind, and build an open and inclusive regional architecture, giving full play to complementarity between countries and institutions. Of course, countries outside the region are expected to play a positive role in East Asian cooperation, rather than the opposite.

Recently, many keep a close eye on the interaction between China and the US in East Asia and the Asia-Pacific. Some are worried if the two can peacefully get along with each other. History tells us that China-US cooperation and the flourishing of the Asia-Pacific go hand in hand. A sound and growing China-US relationship is of great importance to both countries, the Asia-Pacific region and the world at large. China has never done harm to American interests, and we hope the US will not harm our interests. Given the differences in history, culture, development stage and social system, it is only to be expected that China and the US may have divergent views on some regional affairs, colliding interests, or different policies and strategies. But China and the US have also become cooperation partners, who are highly interdependent economically and mutually indispensable in tackling financial crisis and regional issues. We believe that China-US relations are mature enough. The two sides engaged in candid and in-depth communication in recent years over regional situation and issues, as well as on Asia-Pacific policies through meetings at the leadership level, the Strategic and Economic Dialogues, the Strategic and Security Dialogue and consultations on Asia-Pacific affairs. We believe that when both countries work together and take a realistic and cool-headed approach in dealing with disputes, we will be able to find a model of positive interaction in the Asia-Pacific, characterized by peaceful coexistence, healthy competition and win-win cooperation. This will be a conducive force for East Asian cooperation.

In his book *The Next Asia*, the economist Professor Stephen Roach said that the potential of Asia lies not in the simple combination of the development of certain countries, but in the benefits brought about by the integration of Pan-Asia region. This is not only the recognition of achievements in East Asian cooperation in the past 20 years, but also useful advice for future development. We strongly believe that, as long as countries in the region stick to the principles and follow the trend of the times, properly resolve disputes and conflict of interests, and work to maximize common interests, East Asian cooperation will make even greater achievements in the decades to come.

As an important member of East Asia, China has unwavering resolve to pursue a path of peaceful development. China's policy of seeking friendship and partnership with its neighbors will not change, and China's support to ASEAN integration and commitment to East Asian cooperation will not wane. With concerted efforts from all parties, we believe the East Asia in the 21st century will be one of peace and win-win cooperation.

Committing to a Deeper China-ASEAN Strategic Partnership

Remarks at the seminar in commemoration of the tenth anniversary of ASEAN-China Strategic Partnership
July 12, 2013

This year marks the tenth anniversary of the China-ASEAN Strategic Partnership, an important year that links the past with the future of our relations. This seminar, aiming at taking stock of the successful experience in developing our relations, and also planning for future cooperation, is well-timed and highly significant.

In retrospect, China-ASEAN relations have gone through an extraordinary journey, and achieved leapfrog development.

First, we have elevated our friendship and mutual trust to a new high. China established dialogue relations with ASEAN 22 years ago. Among all ASEAN dialogue partners, China was the first to join the Treaty of Amity and Cooperation in Southeast Asia, the first to establish strategic partnership with ASEAN, and the first major country to establish a Free Trade Area (FTA) with ASEAN. This has laid a solid political and economic foundation for our future relations. Eleven years ago, China and ASEAN countries signed the Declaration on the Conduct of Parties in the South China Sea, which advances maritime dialogue and cooperation step by step, and vigorously safeguards regional peace and stability.

Second, we have pushed forward development and cooperation to a new level. In the past ten years, China-ASEAN trade volume has grown by nearly six times from US$78.2 billion to US$400 billion. Mutual investment has increased by nearly four times from US$33.2 billion to US$100 billion. The number of mutual visits has soared by about five folds from 3.87 million to 15 million. Throughout history, the interests of China and ASEAN have never been so intertwined and the bilateral relations so interdependent.

Third, we have ushered East Asian cooperation into a new phase. The rapid development of China-ASEAN relations has greatly propelled the process of East Asian integration. In recent years, China and ASEAN, through mutual support and coordination, have pushed forward the parallel development of ASEAN Plus Three, EAS, ARF and other frameworks, formed a regional cooperation framework with East Asian characteristics, and fostered guiding principles and spirit of ASEAN-led and consensus-based East Asian cooperation that is in line with regional realities and accommodates the comfort levels of all parties.

As the international focus shifts to Asia nowadays, East Asia has been put under the spotlight of the world. Sustained stability and development of East Asia is not only the common aspiration of countries in the region, but also the hope for global economic recovery and revitalization.

China is making unremitting efforts to realize the Chinese dream of prosperity, rejuvenation and happiness. In the next five years, China will import US$10 trillion worth of products with outbound investment exceeding US$500 billion and the number of out-going tourists reaching 400 million. The realization of the Chinese dream will bring greater opportunities for the realization of the dreams of Asian countries.

The Chinese government follows the guiding principle of forging friendship and partnership with its neighbors and attaches great importance to developing good-neighborly friendship and cooperation with ASEAN. Foreign Minister Wang Yi visited Southeast Asian countries shortly after he took office, which fully demonstrated our desire to develop relations with ASEAN. During the China-ASEAN Foreign Ministers' Meeting concluded not long ago, ministers reached broad consensus on furthering the strategic partnership between the two sides. Taking ASEAN as a high priority in its neighborhood diplomacy, China has been committed to growing strategic partnership with ASEAN and handling differences with relevant ASEAN countries through friendly consultation. China will continue to support ASEAN unity, development, community building, and its centrality in regional cooperation.

Many hands make light work. The sound development of China-ASEAN relations relies not only on concerted efforts from governments of both sides, but also the strong support from people of all sectors, especially intellectual input from our think tanks. I hope everyone here will ponder on the following questions:

First, how to enhance political mutual trust. In recent years, the mainstream of China-ASEAN relations has been friendship and cooperation. We should be aware, however, that China and ASEAN are growing rapidly, and growth is always accompanied by pain, and can generate conflict of interests, divergence and even misgivings. In addition, the deliberate smears of some external forces have misled public opinion and caused disturbance to China-ASEAN relations. It is important for both sides to persistently promote mutual understanding, diminish misgivings, and consolidate the foundation for trust. We should strengthen interactions at all levels, give full play to the role of the China-ASEAN Leaders' Meeting, Foreign Ministers' Meeting and other intergovernmental mechanisms, and at the same time, promote communication and dialogue of think tanks, media, youth and other sectors of society. Moreover, security is an important area of China-ASEAN cooperation. We need to discuss how to use the ADMM Plus, ARF and other existing frameworks to strengthen security dialogue and cooperation, and make unremitting efforts to build a peaceful, stable and prosperous regional security environment.

Second, how to deepen practical cooperation. Going forward, we need to make great efforts in the following three areas. First, build an upgraded China-ASEAN Free Trade Area, and improve trade, investment and service facilitation. Synergizing the two major markets of China and ASEAN can generate an effect where 1+1 is greater than 2. This will not only be beneficial for sustained Asian economic development, but also inject impetus to the global economy. Second, promote all-round connectivity. ASEAN is promoting the connectivity of infrastructure, regulations, and people-to-people exchanges, with special emphasis on infrastructure connectivity. China has rich experience in this regard, and the two sides have great potential for such cooperation. China and ASEAN are actively discussing the possibility of establishing an Asia infrastructure financing platform to provide financial guarantee for regional infrastructure development. Third, promote maritime cooperation. We should give full play to the role of China-ASEAN Maritime Cooperation Fund, carry out cooperation in areas of fishery base building, port cities cooperation network, maritime satellite application, marine science and technology, disaster prevention and relief, navigation safety, and search and rescue efforts, and make maritime cooperation a new highlight of China-ASEAN cooperation. I look forward to good ideas and suggestions from you.

Third, how to promote people-to-people exchanges. Amity between people holds the key to sound relations among states. Recent years have seen closer ties between our peoples. China has established almost 30 Confucius Institutes and a Chinese Culture Center in ASEAN countries, and 170,000 students are studying in each other's countries. We need to further deepen cooperation in areas of culture, education, tourism and youth, forge an all-round, multilevel people-to-people exchange structure, and pass down China-ASEAN friendship to future generations. China supports the establishment of a Network of ASEAN-China Think Tanks, and welcomes the active participation of scholars and think tanks from ASEAN countries. I hope all experts present here could send positive messages about China-ASEAN relations, advocate our friendly cooperation, and make contributions to the development of the bilateral ties.

Fourth, how to further strengthen communication in international affairs. Despite the fact that Asia is now at the center of the international arena, it still faces various real and potential challenges in development. The most important challenge is how to cope with the impact of the international financial crisis and the downward pressure on regional economy. China and ASEAN countries should strengthen coordination and cooperation within the regional cooperation frameworks such as ASEAN Plus Three, EAS and ARF, and jointly promote regional integration, so as to keep the momentum of regional economy and sustain Asia's role as the engine of global growth.

The two sides can explore cooperation in non-traditional security areas such as climate change and cyber security, and provide more public security products to the region. As members of the developing world, China and ASEAN countries should strengthen communication and coordination in the reform of the global governance system. This would enhance the voice of developing countries in international affairs, and promote global multi-polarity and democracy in international relations.

Fifth, how to deal with sensitive issues properly. In essence, the South China Sea issue concerns disputes over territorial sovereignty and maritime rights between China and some Southeast Asian countries. It is not a problem between China and ASEAN as a whole. The meddling of external forces will not help resolve the problem, but erode the confidence of China and ASEAN in dialogue and cooperation. Maintaining peace and stability in the South China Sea is in the common interests of China and ASEAN countries. China is willing to fully and effectively implement the Declaration on the Conduct of Parties in the South China Sea (DOC), and conduct consultations on a Code of Conduct in the South China Sea (COC) within the DOC framework. China and ASEAN countries have the confidence and capacity to properly handle the South China Sea issue and build the South China Sea into a sea of peace, friendship and cooperation. I hope everyone present can come up with good suggestions on how to manage regional sensitive issues, and foster a favorable environment for regional development and cooperation.

Linked by mountains and rivers, China and ASEAN countries are good neighbors, good friends and good partners. We all belong to one big family. As a cornerstone and an important pillar for the peaceful and stable development of East Asia, the better China-ASEAN relations grow, the brighter the prospects for East Asia will be.

China and ASEAN:
Good-Neighborly Friendship Bears Rich Fruit

Published on the *People's Daily* on August 29, 2013

On August 29, the Special ASEAN-China Foreign Ministers' Meeting was held in Beijing. As we celebrate the tenth anniversary of China-ASEAN Strategic Partnership this year, relations between the two sides have entered a new historical stage, when we need to build on past successes for further progress in the future. The development of the partnership has not only brought tangible benefits to the two sides, but also greatly promoted the building of the ASEAN Community and made important contribution to peace, stability and prosperity in Asia.

Looking back, China-ASEAN relations have traversed an extraordinary journey and achieved leapfrog progress.

Firstly, we have raised our friendship and trust to a new height. China and ASEAN established dialogue relations 22 years ago. Among all ASEAN dialogue partners, China was the first to join the Treaty of Amity and Cooperation in Southeast Asia (TAC), the first to establish strategic partnership with ASEAN, and the first major country to build a Free Trade Area (FTA) with ASEAN. This has laid an important political and economic foundation for future development of two-way relations. China and ASEAN countries also signed the Declaration on the Conduct of Parties in the South China Sea (DOC) 11 years ago, and they have gradually promoted maritime dialogue and cooperation, thus strongly safeguarding regional peace and stability.

Secondly, we have brought our development and cooperation to a new stage. Over the past ten years, trade volume between China and ASEAN has increased by 4.12 times from US$78.2 billion in 2003 to US$400.1 billion in 2012. Mutual investment has grown by 2.01 times from US$33.2 billion to US$100 billion. The number of mutual visits has also increased by 2.88 times from 3.87 million to 15 million. Throughout the history, China and ASEAN have never seen such closely intertwined interests and interdependence like today.

Thirdly, we have elevated East Asian regional cooperation to a new level. The rapid development of China-ASEAN relations has given strong impetus to East Asian integration. Over the years, China and ASEAN have supported and cooperated with each other to promote the parallel development of ASEAN Plus Three (10+3), the East Asia Summit (EAS), the ASEAN Regional Forum (ARF) and other frameworks, jointly formed a regional cooperation structure with East Asian characteristics and fostered guiding principles of ASEAN-led, and consensus-based East Asian cooperation that is in line with regional realities and accommodates the comfort level of all parties.

As the international focus shifts to Asia, East Asia has been put under the spotlight of the world. Continuous stability and development of East Asia represents not only the common aspiration of regional countries, but also the hope for global economic recovery and revitalization. China is making unremitting efforts to realize the Chinese dream for prosperity of the country, national rejuvenation and people's well-being. In the next five years, China will import US$10 trillion worth of products, make over US$500 billion of outbound investment and see its number of outbound tourists reaching 400 million. The realization of the Chinese dream will bring greater opportunities for the realization of the dreams of other Asian countries.

China follows the policy of forging friendship and partnership with its neighbors and values its good-neighborly friendship and cooperation with ASEAN. When China's Foreign Minister Wang Yi took office this year, his first overseas visit was to Southeast Asia, which fully demonstrated China's strong aspiration for developing relations with ASEAN. During the China-ASEAN Foreign Ministers' Meeting held in Brunei on 30 June, ministers reached broad consensus on further enhancing the strategic partnership between the two sides. Taking ASEAN as a high priority in its neighborhood diplomacy, China has been committed to growing strategic partnership with ASEAN and handling differences with relevant ASEAN countries through friendly consultations. China will continue to support ASEAN unity, development, community building, and ASEAN centrality in regional cooperation.

Linked by mountains and rivers, China and ASEAN countries are good neighbors, good friends and good partners. Both sides belong to one big family. China-ASEAN relations are the cornerstone and important pillar for the peaceful and stable development of East Asia, and the better the relations develop, the brighter the prospects for East Asia will be. In the next stage, the two sides should focus their efforts on the following aspects:

First, we should further enhance political mutual trust. In recent years, friendship and cooperation have been the mainstream of China-ASEAN relations. But we should also bear in mind that China and ASEAN are growing rapidly, and growth is always accompanied by pains, and there may be differences, divergence and even misgivings. In addition, some external forces with ulterior motives deliberately stir up troubles, in order to mislead public opinion and obstruct the development of China-ASEAN relations. It is important for both sides to persistently promote mutual understanding, reduce misgivings, and consolidate the foundation of trust. We should strengthen interactions at all levels, give full play to the role of China-ASEAN Leaders' Meetings, Foreign Ministers' Meeting and other intergovernmental mechanisms, and at the same time, promote communication and dialogue among think tanks, media, youth and all sectors of society. Moreover, security cooperation is an important component of China-ASEAN cooperation. We need to discuss how to use ASEAN Defense Ministers' Meeting (ADMM) Plus, ARF and other existing frameworks to strengthen security dialogue and cooperation, and make unremitting efforts to build a peaceful, stable and prosperous regional security environment.

Second, we should deepen bilateral practical cooperation. Looking forward, we need to make great efforts in the following three areas. The first is to build an upgraded China-ASEAN FTA, and improve trade, investment and service facilitation. Synergizing the two major markets of China and ASEAN can generate an effect where 1+1 is greater than 2. It would not only benefit sustained economic development in Asia, but also inject new impetus into the global economy. The second is to push forward all-round connectivity. ASEAN is promoting the connectivity of infrastructure, regulations, and people-to-people exchanges, with special emphasis on infrastructure connectivity. China is well-experienced in this field, and both sides have great potentials in cooperation. China and ASEAN are actively discussing establishment of an Asian infrastructure financing platform, to provide financial guarantee for regional infrastructure development. The third is to promote maritime cooperation. We should give full play to the role of China-ASEAN Maritime Cooperation Fund, carry out cooperation in areas of fishery base building, port cities cooperation network, maritime satellite application, maritime science and technology, disaster prevention and relief, navigation safety, and search and rescue, and build maritime cooperation into a new highlight in China-ASEAN cooperation.

Third, we should promote people-to-people exchanges. Amity between people holds the key to sound relations among states. Recent years saw ever closer people-to-people exchanges between the two sides. China has established about 30 Confucius Institutes and a Chinese Cultural Center in ASEAN countries, and 170,000 students are studying in each other's countries. We need to further deepen cooperation in areas of culture, education, tourism and youth, forge an all-round, multilevel people-to-people exchange structure, and pass down China-ASEAN friendship from generation to generation. China supports the establishment of the Network of ASEAN-China Think Tanks, and welcomes the active participation of scholars and think tanks from ASEAN countries. I hope that academic circles on both sides could send out positive messages about China-ASEAN relations, highlight its main theme of friendship and cooperation, and make contribution to its development.

Fourth, we should further strengthen communication in international affairs. Though already at the center of the international stage, Asia still faces various real and potential challenges in its development. The most important challenge is how to cope with the impact of the international financial crisis and the downward pressure on regional economy. China and ASEAN countries should strengthen coordination and cooperation within the regional cooperation frameworks such as 10+3, EAS and ARF, and jointly promote regional integration, so as to maintain Asia's role as the engine for global economy. The two sides can explore cooperation in non-traditional security areas such as climate change and cyber security, and provide more public security products to the region. As developing countries, China and ASEAN countries should also strengthen communication and coordination in the reform of global governance system. This would contribute to enhancing the voice of developing countries in international affairs, and promoting the democratization of international relations.

Last but not least, we should deal with sensitive issues properly. The essence of the South China Sea issue is disputes concerning territorial sovereignty and maritime rights between China and some Southeast Asian countries, which is not a problem for China and ASEAN as a whole. Intervention of external forces will not help solve the problem, and will even erode the confidence between China and ASEAN in dialogue and cooperation. Maintaining peace and stability in the South China Sea is in the common interest of China and ASEAN countries. China is willing to fully and effectively implement the DOC in its entirety, and conduct consultations on the COC within the framework of DOC. China and ASEAN countries have the confidence and ability to properly handle the South China Sea issue and build the South China Sea into a sea of peace, friendship and cooperation.

Many hands make work light. The development of China-ASEAN relations requires not only the policy support from governments of both sides, but also the help from all spectrums of the society. I am confident that with the joint efforts of all of us, China-ASEAN relations will embrace an even brighter future.

Building an Asian Community of Shared Destiny

Published on the *Foreign Affairs* in December 2013

Since the beginning of the new century, the global political and economic landscape has undergone rapid changes as the Asia-Pacific region gained gradually in strategic significance and Asian developing countries rose collectively, and it is widely believed that China, the largest economy in Asia and the second largest in the world since 2010, is already a strong engine driving regional cooperation in Asia and will become a global power in the very near future. At the same time, misgivings do exist as to whether China will stick to peaceful development and build friendship and partnership with its neighbors, or seek hegemony by force and adopt a beggar-thy-neighbor approach.

Recently, the Chinese leaders raised the idea of realizing the Chinese dream, the great renewal of the Chinese nation, and advocated the concept of mankind being a community of shared destiny. At the conference on China's neighborhood diplomacy, they made a strong call for "the awareness of community of shared destiny to strike root in neighboring countries," which has drawn wide attention and applause from the international community. History shows that a major country in the world must be, first and foremost, a major country in the region where it belongs. As China becomes more developed, the outside world, its neighbors in particular, will expect China to play the role and shoulder the responsibilities of a major country. China, that puts the neighboring area on the top of its diplomatic agenda and firmly commits itself to deepening good-neighborly cooperation and promoting mutual benefit and win-win results, will make important contribution to world peace and development.

I China stands side by side with its Asian neighbors through thick and thin

China is a member of Asia, and its development and cultural

traditions are rooted in Asia. China has, in its long history, valued national prosperity, people's well-being and friendly relations with its neighbors. It has never engaged in hegemony or expansion, not even at the height of its strength when its economy accounted for 30 percent of the world's total. Instead, through the land and maritime silk roads that ran through China and its neighbors, China reached out to other ancient civilizations to seek mutual complementarity. During China's Ming Dynasty, Admiral Zheng He led the most powerful fleets of the time on seven expeditions to the Western Pacific and the Indian Ocean, where he visited several dozen countries and regions. Rather than colonial plunder, Zheng He introduced porcelain and tea to the places he visited and followed a policy of "building good relations with neighboring countries through virtue" and "sharing the blessing of peace".[2] In modern times, almost all Asian countries went through the humiliation of colonial oppression. They echoed one another's call, and in the course of pursuing national liberalization, upheld their countries' dignity and the right to self-determination.

Since the founding of the People's Republic, Asia has always been a priority of China's foreign policy, and China's foreign strategy and policy have always started from its neighboring area. Nearly 60 years ago, China, India and Myanmar jointly initiated the Five Principles of Peaceful Coexistence, which have been the guidance for our relations with neighboring countries and evolved into basic norms governing international relations. Over the past more than 30 years since reform and opening-up began, China has, on the basis of its strategic judgment that peace and development represent the main themes of the time, normalized relations with the majority of its neighbors and successfully resolved land border questions left over from history with most neighboring countries through peaceful negotiations. China has also put forward the principle of "shelving disputes and pursuing common development" to address disputes over territory and maritime rights and interests, and consolidated and further developed friendly relations with countries in the neighborhood.

China's development has benefited from peace in its neighborhood

---

2   See *Ming Shi, Zheng He Zhuan* (History of the Ming Dynasty, Biography of Zheng He).

and has promoted stability and prosperity in Asia, spurring and supporting Asia's collective rise and vice versa. The past 30 years saw China pursuing reform and opening-up, and achieving modernization that had taken many countries more than a hundred years to achieve. The same period also witnessed rapid economic growth across Asia. In the Asian financial crisis, China mounted an effective response, leading to enhanced regional cooperation through joint efforts with its neighbors and general development and prosperity of Asia. China and its neighbors also successfully overcame the impact of the international financial crisis which broke out in 2008, and made important contribution to global economic recovery. According to IMF Managing Director, Christine Lagarde, Asia has been the consistent global growth leader, driving an astonishing two thirds of total growth in the five years since the crisis hit.[3]

As it develops, China has been giving back to Asia. Over the past ten years, China and Asia have grown together. China's trade with its neighbors surged from over US$170 billion to over US$1 trillion, registering a growth of more than six-fold. In 2012, China's non-financial direct investment in Asian countries and regions was US$54.93 billion, accounting for 71 percent of China's total outbound investment. China has been an important participant and advocate in all the major regional security mechanisms in Asia and made significant contribution to peace and stability in the region. Launched at China's initiative, the Six-Party Talks on the Korean nuclear issue has played a constructive role in maintaining peace and stability on the Peninsula. By adhering to friendly consultation and cooperation in solving differences, China has safeguarded stability in the South China Sea. China has taken concrete actions in support of the reconstruction and reconciliation process in Afghanistan. In spite of the differences and disputes with some neighbors, China has been working unremittingly with the parties concerned to properly solve the problems, thus maintaining the sound momentum in its relations with neighboring countries.

The more developed China becomes, the closer its relations with the world, the neighborhood in particular, will be, and the more it values a

---

3   Christine Lagarde, "Fulfilling the Asian Dream – Lasting Growth and Shared Prosperity", remarks at Boao Forum for Asia Annual Conference, April 7, 2013.

favorable neighborhood environment. History and reality have proven that China's development helps the development of its neighborhood, and is closely linked with the overall development of Asia.

II China and its neighbors face common historic opportunities for development

The economic and political situation in Asia is undergoing profound changes. The global gravity is shifting to the East, and some people believe that the "Asian hemisphere" is rising and the "Asian century" is dawning. Asia, an ancient land of brilliant civilizations, faces the opportunity of collective rejuvenation in the new era.

First, the growth of Asian economy is expected to lead the world and continue to be the important engine for world economic recovery and growth. The IMF forecast puts the growth of emerging Asian economies at 6.3 percent for this year and 6.5 percent for next year, much higher than the global average. Some research institutes predict that Asia as a whole, with the third largest economy in the world now, will surpass the European Union and North America in the foreseeable future.

Second, peace and development are the main trends in the region. Peace is the basis of all development. Asia does not face any major threat of war at the moment. Asian countries generally value the peace and stability in their region and commit themselves to properly addressing disputes through friendly consultation. The hotspot issues are under effective control. A security order featuring cooperative and common security is emerging in Asia.

Third, regional integration has been upgraded. Driven by economic globalization, Asian economic integration has seen constant improvement in system building. Asia has become the most dynamic region in global free trade cooperation. With the number of free trade agreement growing from 70 in 2002 to nearly 260 in 2013 and progress made in the negotiations on the Regional Comprehensive Economic Partnership (RCEP) and the China-ROK-Japan Free Trade Area, economic interdependence has further increased. Various regional mechanisms with respective priorities in political, economic, security, cultural and other fields complement one another and develop in coordination.

Fourth, the sense of community among Asian countries has become stronger. Deepening win-win cooperation, active cultural exchanges and a strong bond of interests has led to a growing sense of Asian identity and pride among Asian countries. To deepen cooperation for better development has become the consensus of Asian countries. Asian countries have formed a community of shared interests.

Having said that, uncertainties affecting peace and stability in Asia still exist and potential risk of local conflicts have not been removed completely. Hotspot and thorny issues keep flaring up, and disputes over territorial sovereignty and maritime rights and interests continue to interfere in relations among countries from time to time. Economic risks, environmental degradation, terrorism and other non-traditional security challenges remain severe. All these require that Asian countries respond with joint and appropriate actions.

Since modern times, China has never been closer than it is today to achieve national renewal. And China has the same aspirations and efforts as other Asian countries in the pursuit of happiness. There is a broad room for Asian countries to achieve development.

III The major-country foreign policy with Chinese characteristics is first and foremost implemented in China's neighborhood

Since the new session of Chinese government took office in 2013, a series of diplomatic measures were launched, marking the momentous start of a new era of China's diplomacy. Building on the guidelines, principles and fine traditions over 60 years of diplomatic practice, China is now actively exploring a path of major-country diplomacy with Chinese characteristics.[4] Asia, where China makes its home and relies for its development, is also where the major-country diplomacy sets sail. Since the beginning of this year, China's neighborhood diplomacy has made new progress with a number of major events carried out successfully, providing a highlight of China's overall diplomatic work.

China and its Asian neighbors enjoyed frequent high-level contacts in 2013. There were exchanges of high-level visits between China and

---

4   Wang Yi, Exploring the Path of Major-Country Diplomacy with Chinese Characteristics, *Guoji Wenti Yanjiu* (China International Studies), No.4, 2013, p.2.

almost all its neighboring countries. President Xi Jinping visited Russia during his first overseas trip, after which he attended the Boao Forum for Asia Annual Conference. In September, he attended the summit meeting of the Shanghai Cooperation Organization and visited four Central Asian countries, and one month later, he attended the APEC Economic Leaders' Meeting and visited Indonesia and Malaysia. China's relationships with Indonesia and Malaysia were upgraded to comprehensive strategic partnerships, and China established strategic cooperative relationship with Brunei and strategic partnership with Sri Lanka. Premier Li Keqiang visited India and Pakistan on his first overseas trip in May, attended the China-ASEAN Expo in September and met leaders of five ASEAN countries on the sidelines of the Expo, and attended the East Asia Leaders' Meetings and paid visits to Brunei, Thailand and Vietnam in October. Vice Premier Zhang Gaoli visited Singapore and attended the annual meetings of bilateral cooperation mechanisms in October. According to available statistics, in the first ten months of this year, over 100 visits at and above the foreign minister level were exchanged between China and its neighboring countries. It is no exaggeration to say that China and its neighbors have maintained close ties as family members do. In addition, through hosting the China-ASEAN Expo, China-South Asia Expo and Special China-ASEAN Foreign Ministers' Meeting, China engaged the neighboring countries in discussions to explore the prospects of regional cooperation and make relevant plans.

China's business cooperation with other Asian countries was further strengthened. From January to May 2013, total trade between China and East and South Asian countries exceeded US$450 billion, surpassing the sum of China's trade with the European Union and the United States in the same period. China became the largest trading partner of Japan, the ROK, the DPRK, Mongolia, Indonesia, Malaysia, Vietnam and Myanmar. Investment cooperation between China and other Asian countries grew rapidly. In 2013, the China-Singapore Suzhou Industrial Park, the China-Singapore Tianjin Eco-City, the China-Malaysia Qinzhou Industrial Park, and the China-Malaysia Kuantan Industrial Park made steady progress, and a large number of major cooperation projects in China's neighborhood made positive headway. Available statistics show that by mid-2013, China had invested a total of US$260 billion in over 40 major cooperation projects in East and South Asian countries, with each project receiving at least US$100 million in funding. These projects accounted for two-fifths of the total number of China-invested overseas major projects and total value of its overseas investment.

China actively promoted regional common development. China attaches importance to connectivity with its neighboring countries. The important initiatives put forward by President Xi Jinping, i.e., the Silk Road economic belt, the 21st century maritime Silk Road and the Asian infrastructure investment bank have received positive response from neighboring countries. With India and Pakistan, consensus was reached during Premier Li Keqiang's visit to the two countries on building the BCIM economic corridor and the China-Pakistan economic corridor. With ASEAN, President Xi stated China's readiness to sign a treaty on good-neighborliness, friendship and cooperation with ASEAN countries and jointly build a China-ASEAN community of shared destiny. Premier Li Keqiang proposed the "2+7 cooperation framework" for the China-ASEAN relationship and raised a number of specific cooperation initiatives regarding the development of the East Asia Summit. Progress was made in the negotiations on the China-ROK free trade area and the China-Sri Lanka free trade area, and two rounds of negotiations were held on the China-ROK-Japan free trade agreement. China also took an active part in the RCEP negotiation, and the process of economic integration in East Asia was further advanced.

China played an active role in upholding peace and stability in the region. Committed to the new security concept featuring mutual trust, mutual benefit, equality and coordination, China took an active part in solving regional hotspot issues and worked with neighboring countries to forge a peaceful and stable environment in the region on the basis of equality, mutual trust and win-win cooperation. On the Korean nuclear issue, China followed the developments in the situation on the Peninsula closely, worked actively to promote peace and encourage discussions, and made unremitting efforts to stabilize the situation. Interactions on resuming the Six-Party Talks among the relevant parties increased and the situation on the Peninsula eased. On the South China Sea issue, China and ASEAN countries, on the basis of implementing the Declaration on the Conduct of Parties in the South China Sea (DOC), launched the consultation on a Code of Conduct in the South China Sea (COC). China also reached agreement with Brunei on enhancing maritime cooperation and established with Vietnam a working group for consultation on joint maritime development. China's policy of "shelving disputes and seeking common development" won the endorsement and support of most of the claimants in the South China Sea. On the Afghanistan issue, China supported Afghanistan's efforts to promote broad-based and inclusive national reconciliation and urged the relevant parties to earnestly implement their commitments to peace and reconstruction in the country while respecting and accommodating the legitimate concerns of countries in the region. China announced that it would host the Fourth Foreign Ministerial Conference on the Istanbul Process of the Afghanistan issue next year, demonstrating China's constructive role in this regard.

Cooperation between China and its Asian neighbors has brought benefits to countries and people in the region and promoted stability and prosperity in Asia.

IV China will work with its neighbors to build a bright future for an Asian community of shared destiny

China's relations with its neighboring countries in Asia have arrived at a new historic starting point. With mutual interests, China and its Asian neighbors have the common need to build an Asian community of shared destiny in order to realize their dreams of development and lasting prosperity.

As an ancient Chinese book says, "Running a country is very much like farming. It must be kept on mind day and night. Thorough plans must be laid out at the beginning and carried out fully to the end." China will be firmly committed to the path of peaceful development and strive for new progress of win-win cooperation with its neighboring countries. China will stick to its guideline of building partnership and friendship with neighbors, steadfastly follow the policy of building an amicable, safe and prosperous neighborhood, act on the principle of "amity, sincerity, mutual benefit and inclusiveness," treat its neighbors with sincerity and make more friends and partners, so that the neighboring countries can benefit from China's development and China can benefit and get support from the development of its neighboring countries in their joint quest for common prosperity.

Looking ahead to the coming year, China will embrace a new phase of active diplomacy in its neighborhood. It will host the APEC Economic Leaders' Meeting, the summit meeting of the Conference on Interaction and Confidence Building Measures in Asia, the Boao Forum for Asia Annual Conference, the Foreign Ministerial Conference on the Istanbul Process of the Afghanistan issue and other important international meetings. Leaders of many Asian countries will come to China to renew friendship and discuss cooperation. According to the decision made at the Third Plenum of the 18th Central Committee of the Communist Party of China, China will deepen all-round reform to allow the market to play the decisive role in allocating resources. This will bring more opportunities for business cooperation between China and its neighbors. Given their historical and cultural affinity, China and its neighbors will only enjoy even closer people-to-people exchanges. With a foundation built through years of careful cultivation, the relations between China and its Asian neighbors will strike deeper roots and grow stronger.

A growing China and a prosperous Asia will be a blessing for the world. China will integrate its own development with that of the whole Asia and make sure that the Chinese dream and Asia's development vision reinforce each other. An Asian community of shared destiny will pool Asian forces, showcase the Asian wisdom and make new contribution to the great cause of peace and development of mankind.

Asia's Security and China's Responsibility

Speech at the luncheon of the ninth CSCAP Conference
Beijing, December 3, 2013

Talking about Asia, it often stands out for its dynamic economy, and is almost the synonym of economic miracles. Yet for some people, the close economic relations among Asian countries are not matched by amicable political relations. They even believe instability and insecurity is on the rise in Asia. How should we view the security situation in Asia? Is Asia secure?

To be fair, Asia has largely maintained peace and stability without large-scale conflicts after the Cold War. This has served as a solid basis for Asia to engage in economic globalization and focus on development. Relations among Asian countries have also improved.

Asia is seeing thriving regional cooperation. FTA negotiations and connectivity projects are making new progress. Economic integration is picking up speed. Asia has become the biggest driving force for global economic growth. This is the main trend of today's Asia. This is also the way to go for Asia.

Having said that, to be frank, Asia is still faced with many security challenges. Legacies of the Second World War and the Cold War, and territorial and maritime disputes continue to affect Asian security.

There are also growing non-traditional security challenges, such as natural disasters, transnational crimes, cyber security, energy and food security. In addition, there are still attempts in our region to seek absolute security through strengthening military alliances. The trust deficit between some countries remains large.

As globalization deepens, countries are more closely interconnected than ever before, becoming a community of shared future. As a general rule, if every country can manage its own affairs well, that would be a big contribution to regional peace, development and security.

In my opinion, only with openness, inclusiveness, mutual respect and cooperation can countries create a virtuous cycle of security and development, and foster a harmonious and stable environment for regional security.

China put forward a new vision of security in the 1990s as it participated ever deeply in regional security cooperation in response to a drastically changed world. We do not believe in the old security concept based on zero-sum game, military hegemony and power politics. Instead, we hold that countries should work together for a new approach to security based on mutual trust, mutual benefit, equality and collaboration. This can be called the 3C security approach, namely, comprehensive security, cooperative security and common security.

Comprehensive security means that security is multi-faceted and interconnected. It includes not just military security, but also economic security, financial security and food security. Comprehensive and trans-boundary challenges such as natural disasters and cross-border crimes require comprehensive solutions. And no country can cope with them alone.

Cooperative security means that security should be realized through cooperation and equal participation of all relevant parties. Disputes should be solved in a peaceful and cooperative manner.

At the recent EAS, Premier Li Keqiang used chopsticks as an example to show the importance of cooperative security. He said one cannot break the chopsticks if you bundle many of them together. His message is that every country has a responsibility and obligation for regional security.

Common security means security for all. No country should seek absolute security for itself or its own security at the cost of others. Security concerns of different countries differ from each other, and countries should consider the security interests of others while pursuing their own. For instance, true security in Northeast Asia should cover all countries in the region, and can only be achieved with a sub-regional security mechanism under which all relevant countries participate as equal.

In his address to this year's Boao Forum for Asia, President Xi Jinping compared peace to air and sunshine. One hardly notices them, yet none can live without them. The peaceful and stable environment in Asia has not come easily. What should we do to maintain and promote security in Asia? Following the new vision of security, China believes that the efforts should be made in the following areas.

Firstly, promoting regional economic integration is the foundation for

Asian security. Development and security are mutually reinforcing. We cannot achieve one without the other. For many countries, development is also the biggest security interest.

Asia has a sound framework for economic cooperation. We have booming 10+1 and 10+3 cooperation. RCEP and China-Japan-ROK FTA negotiations are making progress while financial, connectivity and other cooperation advance in solid strides. We should continue to strengthen the bond of shared interests among countries through regional cooperation, and make efforts for the common development and prosperity of Asia.

Secondly, good relations among major countries are the fundamental guarantee for Asian security and a crucial factor for regional peace and development. They should be rational in judging each other's strategic motives, discard the cold war mentality, respect each other's interests and concerns, and work together to tackle global challenges. These are the expectations of regional countries as well as the responsibilities of the major countries.

Thirdly, existing regional mechanisms provide an important channel for promoting Asian security. We should adhere to multilateralism and oppose unilateralism. Regional mechanisms such as ARF, ADMM Plus and EAS are highly inclusive and productive in promoting practical cooperation in non-traditional security. These increasingly valued and recognized mechanisms should enhance their own institutional buildup and play a bigger role in promoting regional non-traditional security cooperation.

Fourthly, fostering new security architecture is a necessary part of promoting Asian security. There is a growing awareness that security cooperation in our region has lagged far behind economic cooperation. This is not in the interest of Asia's long term development. A regional security architecture that works well for the region and caters to the needs of all parties should be established.

Russia and Indonesia came up with proposals such as signing a principle declaration for Asia-Pacific security cooperation and an Indo-Pacific Treaty of Amity and Cooperation. Some scholars also proposed the concept of Consociational Security Order in the Asia-Pacific. All these are useful ideas.

China believes that the new architecture should be based on the new security vision. It should be conducive to both economic and security cooperation in the region. Naturally, setting up this framework would be an incremental process. It should follow principles such as consensus, non-interference and accommodating the comfort level of all parties. We could start with functional cooperation, so that parties can accumulate mutual trust and raise the comfort level.

In recent years, the development of China has become an important factor and the focus of attention in the evolving situation in the region. How will China use its growing strength, and what role will China play in Asia?

Many of you come from China's neighboring countries and have a good understanding of China's domestic and foreign policies. China still sees itself as a developing country with a busy agenda of conducting reform, achieving development and maintaining stability. For many years to come, the issue at the very top of China's policy agenda remains achieving its own development. Our focus will be on implementing the program of reform, opening-up and development drawn up at the recent 3rd Plenum of the 18th CPC Central Committee, to build a moderately prosperous society in an all-round way for the 1.3 billion Chinese people and realize the two centenary goals.

China has achieved development under the current international order. To keep the order stable, and renew and reform it gradually serves China's interests as well as those of other stakeholders in the region.

At a conference on neighborhood diplomacy not long ago, President Xi Jinping used four phrases to describe China's neighboring policy. They are: amity, sincerity, mutual benefit and inclusiveness. He reiterated that China remains committed to developing friendship and partnerships with its neighbors, and a peaceful path of development to bring more benefits to our neighbors.

China will continue to deepen economic, trade and people-to-people links with other Asian countries. In 2012, China's FDI in Asia amounted to nearly US$55 billion, accounting for more than 70 percent of China's total overseas investment.

This year, Chinese leaders have made many proposals on major cooperation projects during their frequent visits to the neighboring countries, such as, among others, building the Silk Road Economic Belt and the 21st Century Maritime Silk Road, the establishment of an Asia Infrastructure Investment Bank, advancing the 2+7 cooperation framework to enhance China-ASEAN strategic partnership, developing the Bangladesh-China-India-Myanmar Economic Corridor and the China-Pakistan Economic Corridor. These initiatives have been endorsed by other Asian countries. The economic integration of Asia is set to enter a new era of major development, and China will only play a bigger and more active role in it.

China is firmly committed to building a new type of major country relationship. Russia was the first country President Xi Jinping visited after he took office. The two sides reaffirmed their endeavors to achieve a secure and sustainable future for the Asia-Pacific region. The China-Russia comprehensive strategic partnership has set an example for good relations between major countries.

China and the United States have broad common interests in maintaining development and security in the Asia-Pacific. The two countries agreed to build a new type of major country relationship, committing themselves to no conflict or confrontation, mutual respect for core interests and major concerns, and broader practical cooperation for peace, stability and development of the Asia-Pacific and the world at large.

Naturally, such a new type of relationship will not be plain sailing. We owe it to ourselves and to the region to foster mutual trust, engage in sincere cooperation, and avoid the historical trap of major power conflict. This not only serves the interests of the two countries, but also conforms to expectations of regional countries. Tomorrow, Vice President Biden of the United States will start his visit to China, during which leaders of both sides will exchange views on a wide range of topics including bilateral ties, and cooperation in the Asia-Pacific and beyond. We hope this visit will contribute to the building of the new type of major country relationship between China and the United States.

China will continue to firmly support ASEAN community building and ASEAN centrality in regional cooperation. China is the first outside country to sign the Treaty of Amity and Cooperation in Southeast Asia, and the first major country to establish a strategic partnership with ASEAN. During his visit to Southeast Asia, President Xi Jinping stated China's readiness to conclude a China-ASEAN Treaty of Good-neighborliness and Cooperation and build a China-ASEAN community of shared future. This is a strategic initiative by China to upgrade the China-ASEAN partnership under the new circumstances. We look forward to start relevant discussions at an early date.

China will continue to properly handle disputes over territory and maritime rights and interests with relevant countries. Our position of upholding peace and stability in the South China Sea and working for negotiated solutions with countries directly involved remains unchanged. China and ASEAN countries are making joint efforts to implement the DOC comprehensively and effectively and will push forward COC discussions in a positive and prudent manner.

We stand for shelving disputes and seeking common development before the disputes are settled. And we have reached preliminary understanding on common development with some countries. We have also earmarked RMB 3 billion to set up the China-ASEAN Maritime Cooperation Fund, and the first batch of cooperation projects have been planned. On issues of territorial sovereignty and maritime interests, China does not believe in provoking others. Nor would we allow provocation against China's principles and bottom line.

On the issue of Diaoyu Islands, China's activities in the area is the legitimate exercise of its jurisdiction on these islands and should not be seen as an attempt to change the status quo. China's establishment of the Air Defense Identification Zone (ADIZ) in the East China Sea is consistent with international law and international practice. More than 20 countries including the United States, Japan and the ROK have established their own ADIZs since the 1950s. As for the issue of aviation safety in the overlapping ADIZs, China and Japan can and should strengthen dialogue and communication to ensure aviation safety and avoid accidents. We hope that relevant countries will not read too much and not overreact to China's ADIZ in the East China Sea.

China will continue to work for the solution of hotspot issues in Asia. We will firmly push forward the denuclearization of the Korean Peninsula. We sincerely hope that the relevant parties will move in the same direction and make efforts for early resumption of the Six-Party Talks to bring the issue back on the track of negotiations.

China supports extensive and inclusive national reconciliation in Afghanistan. We have actively participated in the peace and reconstruction process and regional cooperation concerning Afghanistan. We will host the Fourth Foreign Ministerial Conference of the Istanbul Process on Afghanistan in 2014 to contribute even more to peace and development in Afghanistan and South Asia at large.

China will continue to be an active player and contributor at regional security mechanisms under ASEAN and promote cooperation within relevant frameworks to make progress. China will take on more responsibility for regional and global security and provide more public security goods to Asia and the world.

China is the biggest contributor of peace-keepers among the five Permanent Members of the UN Security Council. Since 2008, China has sent 15 naval fleets to the Gulf of Aden and the West Indian Ocean for escort missions to merchant ships, and half of the ships escorted fly foreign flags. As a main user of sea lanes, China is ready to explore with other countries cooperation on maintaining the security of sea lanes in relevant seas and oceans, and take up responsibilities accordingly.

The recent super typhoon of Haiyan hit the Philippines and cause huge loss of human lives. In addition to assistance provided in cash and in material, China sent medical teams and the Hospital Ship "Peace Ark" to the country to join in the relief effort. The disaster relief in the Philippines once again reminded us of the urgent and important need to put in place a regional mechanism for disaster relief. We are ready to make greater contribution to the capacity building on disaster management in our region. We will work together with Malaysia to host the ARF Disaster Relief Exercise in 2015.

Forging Ahead with Determination, Playing the Main Melody of Asian Cooperation

Comments on Asian Situation and Neighborhood Diplomacy
December 31, 2013

Journalist: In 2013, Asia holds an increasingly important status and exerts larger influence in the international system. Many people believe that this century is bound to be the "Century of Asia". As the Vice Minister of Foreign Affairs in charge of Asian affairs, what are your comments on the current situation in Asia and its development trend?

Liu Zhenmin: For Asia, the year 2013 is a year of development and cooperation. Against the backdrop of an in-depth development of world multi-polarization and economic globalization, the political and economic situation in Asia remains stable in general, and Asia's status and influence over the international affairs have continued to rise. The rise of the Asian countries as a whole has become a prominent factor that influences the trend of the international situation.

In 2013, Asia enjoyed good governance and harmony, and boosts tremendous development potentials and bright cooperation prospects. First, Asia is still the world's major engine for economic growth. Asia's emerging economies continue to maintain relatively rapid growth, and the International Monetary Fund (IMF) forecasts that, their growth rate for this year will reach 6.3 percent, which is more than twice of the average growth rate of the world. Second, the situation in Asia is stable in general. The Asian countries have reached broad consensus that they would pursue peace, stability, development, and properly handle divergences through friendly consultations. Third, regional cooperation in Asia has increasingly deepened. Asia has become the most dynamic region in global free trade cooperation. Its regional economic integration system is improving, and its political and security cooperation are strengthening continuously.

On the other hand, Asia is also faced with some risks and challenges. Constrained by factors such as the external environment and its own economic structural adjustment, the growth rate of Asia's emerging economies has begun to slow down, and the pressure of economic downturn is increasing. There are still uncertainties influencing regional peace and stability. Hotspot issues and difficult problems frequently exist, and non-traditional security challenges are still severe.

At present, all Asian countries are committed to developing the economy, improving people's livelihood and enhancing mutual beneficial cooperation. I believe that, with the joint efforts of peoples from all countries, Asia will continue to create new wonders.

Journalist: The new central leadership has put forward a series of new ideas and new measures for the diplomatic work. China and other Asian countries enjoy intensive high-level mutual visits, and diplomacy in Asia has shown many bright spots. How do you evaluate China's diplomacy in Asia over the past year?

Liu Zhenmin: The year 2013 is an extraordinary year which marked the first and also a productive year for the diplomatic work of the new generation of the central leadership. We have made the neighborhood diplomacy a priority of China's diplomacy more clearly. The neighborhood diplomacy has opened up a new vista and yielded productive results one after another, displaying a bright view of China's diplomacy.

We focus on improving relations with other Asian countries, and have basically had high-level exchanges with almost all Asian countries. President Xi Jinping attended the Boao Forum for Asia Annual Conference 2013 in April, and attended the APEC Informal Leadership Meeting and visited Indonesia and Malaysia in October. Premier Li Keqiang chose India and Pakistan as stops for his first foreign visit, attended the China-ASEAN Expo and met with leaders of five ASEAN member states in September, and attended the East Asian Leaders' Meetings and visited Brunei, Thailand and Vietnam in October. We have upgraded China's relations with Indonesia and Malaysia to the comprehensive strategic partnerships, and promoted cooperative relations with Brunei and Sri Lanka. Chinese Premier and Indian Prime Minister exchanged visits in 2013, the first time the mutual visits are finished in the same year since more than 50 years ago. Chinese Premier and Pakistani Prime Minister exchanged visits within two months. According to incomplete statistics, China has had about 70 mutual visits with the East Asian countries and South Asian countries at the level of foreign minister or above, with talks and meetings in bilateral and multilateral occasions of about 300 people, and has signed over 200 cooperative agreements. We substantially implement the policy of building good neighborly relationships and partnerships with neighboring countries, comprehensively promote cooperation with neighboring countries, and good-neighbor and friendliness has been advanced to a new stage.

To make China's development benefit the neighboring regions better and promote the common development of China and the neighboring countries, we have proposed a series of major cooperation initiatives, including signing the treaty of good-neighborliness and friendly cooperation among China and the ASEAN countries, establishing the Silk Road Economic Belt and the 21st Century Maritime Silk Road, forging the upgraded version of China-ASEAN Free Trade Area, preparing to establish the Asian Infrastructure Investment Bank, and building China-Pakistan Economic Corridor and Bangladesh-China-India-Myanmar Economic Corridor. These initiatives are conducive to promoting mutual benefit, win-win results and long-term development of the region, and thus have received positive responses.

We focus on the overall peace and stability in the region, actively participate in the resolution of regional hotspot issues, and are committed to working with neighboring countries to jointly create the regional environment with peace, stability, equality, mutual trust, and cooperation for win-win results. We vigorously promote exchanges and cooperation with neighboring countries in fields such as culture, education, youth-to-youth communication, media and tourism, which have improved mutual understanding and friendship with neighboring countries.

As a member of the big Asian family, China share weal and woe with other Asian countries, and form an inseparable community of shared future. China will, as always, adhere to the policy of bringing harmony, security and prosperity to the neighbors, seeking stability and create prosperity together with other Asian countries.

Journalist: The Conference on the Diplomatic Work with

Neighboring Countries set the strategic goals, basic principles and overall arrangement for China's neighborhood diplomacy for the next five to ten years, and clarified the working thought and implementation plans. How will the Ministry of Foreign Affairs actualize the spirits of the Conference and open up a new situation for neighborhood diplomacy?

Liu Zhenmin: The Conference on the Diplomatic Work with Neighboring Countries is an important conference held by the Central Committee of the Communist Party of China (CPC) to coordinate both domestic and international situations and open up a new vista for the neighborhood diplomacy under the new circumstances. It set strategic goals, basic principles and overall arrangement for China's neighborhood diplomacy for the next five to ten years, and opened broader prospects for China's neighborhood diplomacy. With a high level and large scale, the Conference demonstrated the importance the CPC Central Committee attaches to developing good-neighborly relations with neighboring countries and creating a favorable neighborhood environment.

The Conference pointed out that the strategic goals of China's neighborhood diplomacy are to obey and serve the realization of the two "Centenary Goals" and achieve the great rejuvenation of the Chinese nation; to comprehensively develop relations with neighboring countries, cement good-neighborliness and deepen mutually beneficial cooperation; to safeguard and make good use of the important period of strategic opportunities for China's development and uphold national sovereignty, security and development interests; and to strive to grow more friendly political relations, more secure economic ties, deeper security cooperation and closer people-to-people and cultural contacts with neighboring countries. The major policies and principles proposed at the Conference are strategic guidelines for the neighborhood diplomacy in the next stage.

We will have in-depth understanding of and effectively implement the spirits of the Conference on the Diplomatic Work with Neighboring Countries, strengthen communication and exchanges, promote understanding and friendship, and safeguard peace and stability in the neighboring regions; actively participate in regional economic cooperation, accelerate interconnection construction, and promote mutual benefits and win-win results; shoulder the responsibility as a major power, advocate the idea of all-round security, common security and cooperation security, and push forward regional security cooperation; expand non-governmental diplomacy and people-to-people and cultural exchanges, make more friends and form friendship, and spare no effort to create a community of shared future among China and the neighboring countries.

Journalist: In 2013, the economy of the Asian countries continued to lead the global economy, but risks and challenges were on the rise, and the pressure of maintaining sustainable development builds up. As the largest economy in Asia, what role can China play in promoting sustained momentum of economic growth and carrying out mutually beneficial cooperation in Asia?

Liu Zhenmin: Asia has maintained relatively rapid economic growth in 2013, and is still one of the regions with the most vigor and potential in the world today, serving as a major engine to stimulate the world's economic recovery and growth. The IMF forecasts that, the economic growth rates of Asia's emerging economies in this year and the following year are 6.3 percent and 6.5 percent respectively, much higher than the world's average. Meanwhile, affected by the adjustment of monetary policies in developed countries, the market fluctuation in some countries is aggravated, and the risk of economic downturn in Asia increases. Although China's economy is also faced with some pressure, China is still a major power with the highest growth rate in the world, playing a crucial role in stabilizing regional economy.

With very close economic and trade relations with other Asian countries, China has become the largest trade partner, largest export market and an important investment source of many Asian countries. China carries out pragmatic cooperation widely with other countries in the region, implements cooperation actively with other Asian countries in fields such as high and new technology, new energy, energy conservation and environment protection, and promotes cooperative projects of industrial parks, ecological parks and economic and trade zones with neighboring countries. The total trade volume among China and the East Asian and South Asian countries from this January to September exceeded US$834.3 billion, which took up 27.3 percent of

China's total foreign trade volume during the same period. About 70 percent of China's foreign investment goes to the Asian countries and regions. It is estimated that the rate of contribution of China's economic growth to Asian economic growth has surpassed 50 percent.

China actively promotes regional economic integration. The significant cooperative initiatives put forward by China, such as upgrading the China-ASEAN Free Trade Area, and establishing the Asian currency stability system, Asian credit system and Asian investment and finance cooperation system, have shown China's active will in deepening regional cooperation and promoting mutual benefits and win-win results. China actively boosts interconnection, and puts forward a series of major initiatives, such as building the Silk Road Economic Belt, the 21st Century Maritime Silk Road, the China-Pakistan Economic Corridor, and the Bangladesh-China-India-Myanmar Economic Corridor, as well as preparing to establish the Asia Infrastructure Investment Bank. With important willingness and eye-catching initiatives, the implementation serves as the key. We will strive to turn the strategic concept of regional cooperation into tangible results that benefit the neighboring countries, and the peoples of China and other Asian countries.

The world and especially the Asian countries, pay close attention to China's future development. The decision concerning comprehensively deepening reforms was adopted at the third Plenary Session of the 18th CPC Central Committee. These reform measures will help China's economy maintain the momentum of sound development. The domestic demand, especially consumer demand, will expand continuously, while foreign investment will increase substantially. These will bring more opportunities to the cooperation among China and other Asian countries. China will continue to make contributions to Asia's development and prosperity.

Journalist: Since coming into power, Japan's Shinzo Abe government has continued to hold tough policies toward China, and Abe paid homage to the Yasukuni Shrine a few days ago. Meanwhile, the Diaoyu Islands issue remains deadlocked. How do you view the future development of China-Japan relations?

Liu Zhenmin: The China-Japan relations have been beset by continuous, serious difficulties since the Japanese side created the farce of "purchasing" the Diaoyu Islands last year. On December 26, Japan's Prime Minister Shinzo Abe, going against the tide of history and openly provoking international justice and human conscience, blatantly paid homage to the Yasukuni Shrine where class-A war criminals of the World War II are consecrated. It erected a new and major political barrier to the China-Japan relations which have already been beset by severe difficulties.

The Abe government strives to push a radical right conservative path with historical revisionism as its core, leading Japan to a very dangerous direction and bringing about increasingly serious negative effects to the security of East Asia. The neighboring countries in Asia and international community should be of high vigilance to resolutely prevent the Japanese right-wing forces from turning the clock back to the backtrack of the history, so as to maintain the hard-won overall peace and stability in East Asia.

We have always advocated developing long-term, sound and stable China-Japan relations on the basis of the four political documents between China and Japan in the spirit of "taking history as a mirror and looking into the future." It is imperative that Japan correct its mistakes, stop provocative words and deeds, and make practical efforts to improve the relations between the two countries.

Journalist: Some uncertain factors exist in the security situation in Asia. The DPRK nuclear and other issues are still in stalemate and the future direction of the situation has considerable uncertainty. How will China play its role in maintaining regional peace and stability?

Liu Zhenmin: The overall situation in the Asia-Pacific region is stable, but some hotspot and sensitive issues also exist. China pays high attention to the development of the Korean Peninsula situation and has always been committed to actively promoting peaceful discussions so as to push forward the proper resolution of the Korean Peninsula nuclear issue and to making unremitting efforts to maintain peace and stability on the Peninsula. It is China's unswerving position to realize denuclearization of the Korean Peninsula, maintain peace and stability on the Peninsula, and resolve the issues peacefully through dialogues and consultations. The DPRK nuclear issue is a long-standing and intricate issue. We hope that all relevant parties meet each other halfway, do more things that can enhance trust and ease the situation, and return on the track of dialogue at an early date.

Journalist: In 2013, the situation in South China Sea was obviously eased, and the maritime cooperation among China and the neighboring countries stepped into a new stage. Please brief us on China's efforts to promote maritime cooperation and relevant achievements.

Liu Zhenmin: Our position has always been that China has indisputable sovereignty over the Nansha Islands and the adjacent waters. The relevant parties' disputes over some islands and reefs of the Nansha Islands and the delimitation of some waters in the South China Sea should be solved properly and peacefully through negotiations by parties directly concerned. Before the settlement of disputes, it is advisable to "shelve disputes and carry out common development." To this end, we have made active efforts. In 2013, along with the ASEAN countries, we actively implemented the Declaration on the Conduct of Parties in the South China Sea (DOC) and launched the consultation on the formulation of the Code of Conduct in the South China Sea (COC) within the DOC framework. China has reached important consensuses with Brunei and Vietnam respectively, taking a big step forward on the issue of jointly developing the South China Sea.

Promoting maritime cooperation is in line with the common interests of China and the ASEAN countries, and is also the most effective way to resolve differences and ease conflicts. It is helpful for all parties to accumulate trust and create conditions and atmosphere for the final settlement of the disputes. China is committed to promoting maritime cooperation with the ASEAN countries. During his visit to Indonesia, President Xi Jinping proposed the initiative of building the 21st century "Maritime Silk Road" together with the ASEAN countries for the purpose of further expanding maritime cooperation with the ASEAN countries and promoting common development. We have communicated and coordinated closely with the ASEAN countries, made full use of the China-ASEAN Maritime Cooperation Fund set up by the Chinese government, and determined on a number of maritime cooperation projects. We are willing to work with the ASEAN countries to make the maritime cooperation a new highlight of China-ASEAN mutually beneficial cooperation, benefit peoples of all countries and make the ocean a friendly tie linking China and the Southeast Asian countries.

Journalist: In 2013, China's neighborhood diplomacy has yielded fruitful results, yet the tasks for the next year are even heavier. Please look forward and brief us on the direction of the diplomatic work in Asia in 2014.

Liu Zhenmin: The year 2014 will be the beginning year to implement the spirit of the Conference on the Diplomatic Work with Neighboring Countries, and also a year requiring diligent work to carry out the major cooperation initiatives advanced by the state leaders. The diplomatic work in Asia will be more arduous. We will continue our commitment to building good neighborly relationships and partnerships with neighboring countries and bringing harmony, security and prosperity to the neighbors, actively practicing the new concept of "closeness, sincerity, sharing in prosperity, and inclusiveness" in neighborhood diplomacy, further enhancing political mutual trust with other Asian countries, expanding convergence of interests, strengthening exchanges and cooperation, managing conflicts and disagreements, and jointly fostering a peaceful, stable, cooperative and prosperous regional environment.

We will further deepen the good neighborly and friendly relations with other Asian countries. China will host significant international conferences including the APEC Informal Leadership Meeting, the Conference on Interaction and Confidence-Building Measures in Asia, the Annual Conference of Boao Forum for Asia and the Foreign Ministerial Conference on the Istanbul Process of the Afghanistan issue. We will make use of these platforms to renew our friendship and talk about our cooperation with the leaders of other Asian countries. We will adhere to the new security perspective of mutual trust, mutual benefit, equality and cooperation, play a constructive role more actively in regional hotspot issues, manage disagreements with relevant countries, and maintain regional peace and stability.

We will further expand the convergence of interests with other Asian countries. China will continue to implement the opening-up strategy of mutual benefit and win-win results, vigorously push forward pragmatic cooperation with other Asian countries, speed up connectivity development and maritime cooperation, and promote the building of the Silk Road Economic Belt and the 21st century Maritime Silk Road. We will actively participate in regional cooperation, establish an upgraded version of the China-ASEAN Free Trade Area, and implement the initiatives concerning China-ASEAN "2+7" cooperation framework, China-Pakistan Economic Corridor, and Bangladesh-China-India-Myanmar Economic Corridor. We will push forward the negotiations on free trade agreements (FTAs) such as the Regional Comprehensive Economic Partnership (RCEP), China-ROK FTA and China-Japan-ROK FTA, so as to continuously promote regional economic integration.

We will further strengthen people-to-people and cultural exchanges with other Asian countries. The year 2014 is the year of Friendly Exchange between China and the ASEAN, Mongolia, India and other countries. China, India and Myanmar will commemorate together the 60th anniversary of the release of the Five Principles of Peaceful Coexistence. We will explore various channels to expand personnel exchanges, enhance the development of Chinese Culture Centers and Confucius Institutes (Classes), promote understanding and friendship among the peoples, expand and deepen exchanges and cooperation with other countries in fields such as education, culture, tourism, media and think tanks, and compose a new chapter of people-to-people and cultural exchanges.

The year 2014 will be a promising year for China's neighborhood diplomacy. We will follow the guidance of the spirits of the third Plenary Session of the 18th CPC Central Committee and the Conference on the Diplomatic Work with Neighboring Countries, face up to the difficulties, forge ahead with determination, and do all the work well in neighborhood diplomacy with great efforts, so as to well maintain national sovereignty, security and development interests, better serve the interests of the motherland and the people, and make greater contributions to the stability and development in Asia.

Committing to Win-win Cooperation and Forging the Asian Community of Shared Future Together

Published on the *International Studies* in February 2014

Since the beginning of the 21st century, the rise of Asian developing countries as a whole has greatly reshaped the political and economic landscapes of the world. As an Asian country endeavoring to realize its dream of national rejuvenation, China has also proposed to cooperate fully with its Asian neighbors to jointly build peace, stability and prosperity in the region.

Since the Communist Party of China (CPC) proposed to "raise awareness about mankind sharing a community of shared future" in its Report of the 18th CPC National Congress, Chinese leaders have elaborated on the concept of "community of shared future" on many occasions and promoted all-round international and regional cooperation. China held an unprecedented conference on the diplomatic work with neighboring countries, proposing to make the idea of "community of shared future" take root in the neighboring countries.

It is fair to say that building a "community of shared future" has become a banner in innovating China's diplomatic theories and practices in the new era, and building such a community in Asia has become the direction for China's neighborhood diplomacy.

Much can be expected as China's foreign policy ascends to a new height.

I The meaning of the Asian community of shared future

Despite many conflicts and disputes, peace, development and cooperation are the mainstreams of today's world, and economic globalization and regional integration are the dominant trends. China has never been so close to the center of the world stage like today, neither has it been so closely connected with the destiny of the outside world. By proposing the "community of shared future," China is putting forward its plan for the future welfare of Asia and even the world with a focus on the mighty cause of human development in the era of globalization and China's long-term development and the prosperity and stability of its neighboring countries. When advocating the building of an Asian community of shared future, China draws upon advanced experiences from the West, builds on regional integration and the consensus among regional countries for common development and security, and emphasizes handling state-to-state relations properly with the sense of the community of shared future. In short, the country is setting a higher objective with more profound theoretical connotations.

First, common development is the core essence. Developing the economy and improving people's livelihood are the core tasks all nations face. Against the backdrop of economic globalization, interests of all the countries are deeply intertwined and bound together for good or ill, making common development the only way out. While pursuing their own development, nations should also work to connect their development strategies with the process of regional integration, and take other's legitimate concerns into consideration while safeguarding their own interests to draw on each other's advantages and promote win-win cooperation.

Second, to safeguard the security environment through mutual trust and coordination. A peaceful and stable international and regional environment is the precondition for all nations to pursue common development, which requires all parties to make joint efforts and take up the responsibility of maintaining regional peace and stability together. Under the new circumstances, it is better to follow the new security vision featuring mutual trust, mutual benefit, equality and coordination, discard the mentality of Cold War and zero-sum game to achieve comprehensive, cooperative and common security.

Third, to advance institutional buildup with openness and inclusiveness. Asia is a highly diversified region with different nations following different paths of development. The social system and development path chosen by the countries themselves should be respected, and regional diversity could be turned into complementary vigor and stimulus for development. Asia should continue learning advanced experiences from other regions of the world and promote the various cooperative mechanisms to play their roles in a coordinated manner. Meanwhile, non-Asian countries should be welcomed to participate in Asian cooperation and play constructive roles for stability and development in the region. All sides should commit to resolving conflicts and disagreements through dialogues and consultations and build a new regional security structure for lasting peace and tranquility in Asia.

Fourth, to build consensus through cultural exchanges. All nations in Asia have their respective great cultural traditions, and the diversified cultures coexist in the globalized world. The mentality of clash of civilizations should be discarded and cultural exchanges and convergence should be encouraged. Asian countries should carry forward and spread the cultural tradition of valuing good faith and harmony, adhere to the Asian way of mutual respect, reaching consensus through consultation and accommodating each other's comfort level, and encourage cultural exchanges and mutual learning to ensure harmonious coexistence.

Fifth, to strengthen bonds through mutual assistance in times of difficulty. Helping and looking out for one another has been the valuable asset shared by Asian countries. In the globalized world today, various global challenges arise frequently, such as international financial crises, natural disasters and climate change. No country can survive and flourish on its own. Global challenges require countries to cooperate in response, and the act of supporting the poor and assisting the weak should be encouraged. The right way should be the strong supporting the weak and the rich helping the poor in the period of ups and downs.

Building the Asian community of shared future is a comprehensive and systematic project covering the political, economic, security, social and cultural fields. It is linked with the interests and development of all Asian nations and will determine how the rising Asia would interact with the world and what they would communicate. Therefore, its connotation and denotation will be enriched and expanded continuously with the passing of time.

II Building the Asian community of shared future is the inevitable option of historical development

Advocating the building of the Asian community of shared future is the strategic option made by China based on its own development needs. It conforms to the law of historical development and is in line with the common aspiration and interests of Asian countries. Looking back at the trajectory of historical development, it is not difficult to come to the conclusion that building the Asian community of shared future enjoys a profound historical heritage, reflects the distinct features of the times and is highly relevant to the present world.

(1) The Asian community of shared future captures the best of Asian traditions, history and cultures. Long being ahead of the rest of the world, Asian cultures have made great contributions to the progress of human civilization. Linked by mountains and rivers and having a long history of engagement with each other, Asian countries have formed unique historical, cultural, economic and social connections. The land Silk Road and the maritime Silk Road, which both originated from China, have become the bridges of friendship carrying economic, trade and cultural exchanges among Asian countries. Over 600 years ago, General Zheng He led the most powerful fleet in the world then in seven voyages to over 30 countries and regions around the globe. What he implemented was a peaceful foreign policy of "promoting good neighborliness by virtue," exporting porcelain, silk and tea instead of colonial exploitation. According to statistics collected by Western scholars, Asia's economic size accounted for two thirds of the world's total till the beginning of the Industrial Revolution. In spite of the wars in Asian history, peace and cooperation have always been the solid foundation for Asia's prosperity.

In the modern times, however, Asia missed out on the opportunity of the Industrial Revolution and was bullied for lagging behind, suffering from Western invasion and colonization. Echoing with each other, Asian countries engaged in the great cause of national liberalization. Many social elites started searching for the identities of their own countries and Asia from the perspective of modern international relations and exploring ways of cooperation for Asian countries. Asia suffered tremendously from the two World Wars and the Japanese fascism left unprecedentedly painful memories to the people in Asia. The "Great East Asia Co-Prosperity Sphere" beautifying hegemony and aggression was thrown into the dustbin of history. After the end of WWII, developing countries in Asia and Africa held high the banner of regional collaboration, and the Bandung Conference greatly encouraged Asian nations to unite for self-strength.

However, since Asia was at the frontline of the Cold War and was split into two different camps – the East and the West Blocs, regional cooperation here did not actually start until the end of the Cold War.

Historical experiences show that Asian countries benefit from cooperation and lose if they engage in conflicts. Asian rejuvenation has not come by easily, and uniting for self-strength is the right path to pursue.

(2) Building the Asian community of shared future conforms to the trends of global and regional development. As multi-polarization, economic globalization and social informatization grow in-depth, humanities are living in one global village today. Interests of all nations are intertwined and they share the same future. With the continuous readjustment of the international political and economic orders, the long-term development of the world could never be built on the basis of a few amassing vast amounts of wealth while others remain impoverished. Common development should be the only option. Europe has led the world in terms of regional integration, and North America, Latin America and Africa have promoted regional integration vigorously, making it a general trend for Asia to deepen regional cooperation.

After six decades since the end of WWII, Asian nations have overcome various risks and challenges, shaken off the shackles of poverty and weakness, and created the universally acknowledged "Asian miracle" to become a region with the most rapid development and the greatest potential for growth. According to the estimation by the International Monetary Fund (IMF), Asia has contributed two thirds of the world's economic growth in the five years since the outbreak of the international financial crisis in 2008.

In the meantime, regional cooperation has provided strong impetus for Asia's development, further increasing interdependence among Asian nations. In the new century, intra-regional trade in Asia has increased from US$800 billion to over US$3 trillion and the trading interdependence has reached over 50 percent. According to the report of the Asian Development Bank, Asia has become the most dynamic area in the world in the building of free trade areas, with the number of free trade agreements soaring from over 70 in 2002 to over 250 in early 2013. Negotiations on the China-ROK and China-Japan-ROK free trade areas and the Regional Comprehensive Economic Partnership (RCEP) have made new progresses. ASEAN-led East Asian regional cooperation mechanisms have strongly driven the development of the region. It is the common aspiration of Asian countries to build the Asian community of shared future, deepen the convergence of their national interests and realize common development.

(3) Building the Asian community of shared future reflects the common interests and pursuits of Asian countries. Peace and development are not only the general trends in Asia but they also reflect the shared interests and the popular will of all countries here.

Most Asian countries are developing countries, for which development is still the key to resolving major conflicts and problems while developing the economy and improving people's livelihoods are their top priorities. Asia is widely recognized as the most dynamic region with the greatest development potential throughout the world. However, its external environment is not promising due to the lingering profound impact of the international financial crisis and the unsteady recovery of the world economy.

With the US leading the negotiations on building the Trans-Pacific Partnership (TPP) and the US and the EU initiating talks on the Transatlantic Trade and Investment Partnership (TTIP), the global landscape of trade is going through significant readjustments, forcing Asia to make more efforts to become a stable third pole in the world economic order.

Since Asian countries are at different levels of development and confronted with the heavy task of domestic economic reform, they will find it more difficult to promote discussions on trade liberalization and economic integration. As a result, only by making concerted efforts and advocating the Asian plan together could Asia gain the initiative in global rule-making and win long-term interests.

The rapid rise of Asia has attracted increased strategic input from major powers in the world, which offers tremendous impetus for Asia's development and makes regional cooperation and competition rise simultaneously. Asia is confronted with many troubles and problems in the security field, including issues left over by history like the remnants of WWII, the grievances accumulated over the Cold War years and the maritime disputes as well as non-traditional security threats such as natural disasters, transnational crimes, cyber security and energy and food security. Military alliances in Asia are being strengthened and trust deficits are high among some countries. It is the common aspiration of Asian countries to resolve their conflicts and disagreement by increasing dialogues and consultations and safeguard regional stability based on mutual trust and cooperation and pursue development in a peaceful and favorable environment.

(4) Advocating the Asian community of shared future demonstrates the trend of China's neighborhood diplomacy. After 30 years and more since its reform and opening-up program, China has completed the modernization process which took the Western powers over a century to finish and registered remarkable achievements in development. It is attributed to the peaceful and stable neighboring environment and, in turn, promotes Asia's stability and prosperity. China's development and the rise of Asia as a whole are mutually reinforcing.

China follows a path of peaceful development and implements a policy of good neighborliness and friendship to deepen cooperation with neighboring countries, thus making great contribution to Asia's prosperity and stability. The trade volume between China and East Asian and South Asian countries was over US$1.14 trillion in 2013. Among China's ten largest trading partners, half are from Asia. Almost 70 percent of Chinese foreign investment goes to Asian countries and region. China has become the biggest trading partner, the biggest export market and the major source of investment for many Asian countries. As an important participant and advocate of Asian security frameworks, China has been a staunch force to safeguard regional peace and security in Asia. As China further deepens its reforms in an all-round way, its interests and those of other Asian countries would be ever closely linked.

The rising China has brought about quiet changes in the mentality of regional countries. On the one hand, they hope to share more dividends from China's development and ask China to shoulder more responsibilities; on the other hand, their worries aggravate that China would become the dominant power in Asia, taking development opportunities of the others and embarking upon the old way where a strong nation was bound to seek hegemony.

Historical experiences show that the road of growth starts from its neighboring region. For China to realize its dream of national rejuvenation, it needs, first and foremost, to be recognized and supported by other Asian countries, and the dream of the Chinese people is tied with that of other Asian nations. The more China develops, the closer is China's connection with Asia and the world, and the more it needs to consolidate its strategic reliance on Asia. The Asian community of shared future carries forward and innovates the long-standing good-neighborliness policy with an aim to embed China's development and security in the overall development and common security of Asia. This is not only the responsibility taken and promise made by China for Asia's future, but also the strategic option China makes to achieve its own development in the long run.

III Historical opportunities for building the Asian community of shared future

The development of Asia has been influencing the international political and economic orders in an all-round manner. Many believe that the "Asian hemisphere" is rising, the "Asian century" is coming and the power center of the world is moving eastward. Viewing from the trends of world and Asian development, Asia is bestowed with the opportunity of achieving collective rise in the new era.

First, Asia continues to take the lead in the world economy. Asia is expected to be a major engine of the recovery and growth of world economy for a long time to come. Some research institutions estimate that Asia, as the world's third largest economic zone, would surpass the EU and North America in the foreseeable future. According to the forecast of the Asian Development Bank, the share of Asian economy in the world economy will increase from 28 percent in 2010 to 44 percent in 2035 and to 52 percent in 2050. The per capita GDP in Asia by then will reach the current level of Europe.

Second, peace and stability are the trends of the times. Peace is the foundation for development. Asian countries have conflicts and divergences, but they all attach importance to maintaining regional peace and stability and are committed to resolving disagreements properly through friendly consultation and putting hotspot issues under effective control. The regional security situation in Asia is heading increasingly towards cooperative security and common security.

Third, institutions and mechanisms offer the strong support needed. Various cooperation mechanisms have developed rapidly in Asia. In East Asia, a range of cooperation frameworks, such as the ASEAN+3 cooperation, multiple ASEAN+1 arrangements, the China-Japan-ROK dialogue and cooperation, the ASEAN Regional Forum (ARF), the East Asia Summit (EAS) and the Six-Party Talks, have been formed. The South Asian Association for Regional Cooperation (SAARC) covering eight South Asian countries has been developing with a strong momentum. Cooperation frameworks in the political, economic, security, people-to-people and cultural fields complement each other and develop jointly, offering an effective platform for the building of the Asian community of shared future.

Fourth, the sense of Asia has been strengthened. The exchange of Asian countries in the various areas complements each other and the interests converge deeply. As Asian countries are rising robustly, their sense of belonging to Asia and identification with Asia is growing. The key to state-to-state relations lies in the affection between their peoples. For most of the Asian economies, over 80 percent of their inbound tourists are from Asian countries.

The intra-regional interactions and exchanges have cemented an increasingly solid social and popular foundation for building the Asian community of shared future.

However, we are also keenly aware of the historical and practical obstacles. First, Asia is highly diversified. In terms of development level, political system, society, culture and religious belief, Asian countries differ greatly from one another, thus making it more difficult to promote regional integration.

Second, there remain uncertainties in Asia's security situation. Hotspot and sensitive issues intertwine with one another, and issues regarding the Korean Peninsula, the East China Sea, the South China Sea and Afghanistan remain sensitive and complicated. Japanese leaders challenged the results of WWII and the post-war international order by blatantly paying homage to the Yasukuni Shrine where WWII class-A criminals are enshrined, causing tensions in the region.

Third, the building of a regional structure is lagging behind, making it difficult to forge a single governance institution like the EU in the short term. The existing mechanisms for cooperation would coexist in the long run and it is difficult to coordinate the interests of all parties concerned.

Moreover, the "noodle bowl" effect resulting from the many free trade agreements is prominent, and the rivalry in the approaches to promoting economic integration is salient. Security cooperation in Asia has long lagged behind economic cooperation among the countries, and a regional structure for security cooperation suitable to the reality and catering to the needs of all sides has not yet been established. The existence of such difficulties and obstacles reflects the urgency and significance of building the Asian community of shared future. All nations uniting and coordinating to cope with challenges would be an inalienable part of building the Asian community of shared future.

Opportunities and challenges coexist. Looking to the future, we see both unprecedented opportunities and unprecedented challenges. The difficulty in seizing the opportunities, coping with the challenges and turning crises into opportunities is also unprecedented. This requires all nations in Asia to make elaborate plans and strive for the great cause together. We have every reason to be confident with building the Asian community of shared future.

IV China is the major force in building the Asian community of shared future

Building the Asian community of shared future is the common cause of Asian nations. For China, maintaining its own stability and sustained development while promoting good neighborliness and cooperation would inject great vigor to the building of the Asian community of shared future. China should also play the leading and exemplary role, contributing more Chinese wisdom and proposing more Chinese initiatives in regional affairs while promoting other nations to firmly establish the awareness of sharing a community of shared future so that they could take the building of the community as the common goal and take conscious actions for its building.

(1) Promoting good neighborliness through high-level exchanges. The neighborhood where China finds itself is what China depends on for survival, development and prosperity. It is also the place from which the major country diplomacy with Chinese characteristics sets sail. The new Chinese leadership attaches great importance to the neighborhood diplomacy. In 2013, President Xi Jinping, Premier Li Keqiang and other high-level leaders engaged in frequent diplomatic activities and have basically had high-level exchanges with almost all Asian countries. Seeking to deepen relations with other Asian countries in an all-round way, China has forged comprehensive strategic partnerships with Indonesia and Malaysia and upgraded its cooperative relations with Brunei and Sri Lanka. Chinese Premier and Pakistani Prime Minister exchanged visits within two months. Chinese Premier and Indian Prime Minister exchanged visits in 2013, the first time the mutual visits were finished in the same year in over 50 years.

According to the statistics available, China has had about 70 mutual visits with East Asian and South Asian countries at the level of foreign minister or above, with talks and meetings on bilateral and multilateral occasions of about 300 people. For the first time since its founding, China has held the "Conference on the Diplomatic Work with Neighboring Countries," which clarifies the guidelines and general principles for conducting neighborhood diplomacy under the new circumstances.

China will continue its commitment to building good-neighborly relationships and partnerships with its neighbors and bringing harmony, security and prosperity to them, and actively practice the principles of "amity, sincerity, mutual benefit, and inclusiveness" in its neighborhood diplomacy. China will take the lead to cooperate fully with other Asian countries so as to make peace, cooperation and win-win outcome the shared ideas and code of conduct complied by all countries in the region.

(2) Promoting Asia's common development with China's own development. In recent years, China has contributed more than 50 percent of the economic growth of Asia, thus becoming the major stimulator for the overall development of Asia. The third Plenary Session of the 18th CPC Central Committee has laid out an overall plan for deepening reforms comprehensively. The new round of reforms is not only in the interests of China itself but will also promote China's cooperation with foreign countries. China will keep pushing forward its economic transition, explore its market potential and increase investment abroad, bringing new opportunities for the whole world. There is no doubt that Asian countries, enjoying a favorable geographical position, will benefit first.

Following the development trend of Asia and finding the areas where the interests of Asian countries converge, China has put forward a series of major initiatives, including building the Silk Road Economic Belt, the 21st century Maritime Silk Road, the China-Pakistan Economic Corridor, the Bangladesh-China-India-Myanmar Economic Corridor, the China-ASEAN "2+7" cooperation framework, and the Asia Infrastructure Investment Bank. China will work closely with other Asian countries to ensure that these initiatives are implemented and become effective as soon as possible. In that way, stronger impetus would be offered for the economic development of Asia and people in all nations could benefit as well.

Guided by the proper view of righteousness and benefit, China will increase its assistance and support of other Asian countries. We will continue to increase our assistance to other Asian developing countries with no strings attached, especially to the least developed ones, and help them achieve autonomous and sustainable development. We will encourage well-established Chinese enterprises to expand investment in Asian countries and actively take part in their economic development. When Asian countries are struck by such difficulties as natural disasters, China will offer a helping hand as it has done in the past.

(3) Promoting the building of the community of shared future through regional cooperation. China has always actively participated in and promoted regional cooperation. The existing cooperative frameworks in Asia, like the ASEAN, the ASEAN+3, the EAS and the SAARC, have their distinct features and priorities. They should give full play to their own advantages while complementing each other to form a united force. China will continue its firm support for the building of the ASEAN Economic Community (AEC) and ASEAN's leading role in promoting regional cooperation.

China has put forward the China-ASEAN "2+7" cooperation framework and held discussions with ASEAN member states on the signing of treaties of good-neighborliness and friendly cooperation. It is also striving to establish an upgraded version of the China-ASEAN FTA. All these initiatives will help increase China-ASEAN cooperation substantially. China is willing to deepen its practical cooperation with the SAARC by building the China-South Asia Expo as a platform for cooperation.

China is accelerating the implementation of the free trade agreements with its neighbors, actively promoting the comprehensive economic partnership in the region while negotiating on the building of China-ROK, China-Sri Lanka, China-Mongolia and China-Japan-ROK FTAs. We will also strengthen coordination with other Asian countries, and jointly seek to establish the Asian currency stability system, the Asian credit system and Asian investment and finance cooperation system to promote regional economic integration and enhance the capacity of Asian countries in jointly resisting external risks and challenges.

In 2014, China will host many major international conferences, including the APEC Informal Leaders' Meeting, the Conference on Interaction and Confidence-Building Measures in Asia, and the Annual Conference of Boao Forum for Asia. Leaders of the neighboring countries will come to China to attend these conferences, renewing friendship and discussing cooperation with China. The regional cooperative frameworks will be the solid foundation for the building of the Asian community of shared future.

(4) Safeguarding Asian peace and stability through security cooperation. China follows the path of peaceful development and plays its role actively as a responsible major country. It advocates the new security vision of mutual trust, mutual benefit, equality and cooperation while pursuing comprehensive security, cooperative security and common security. China is dedicated to making concerted efforts with its neighbors to create a peaceful and stable regional environment featuring equality, mutual trust and cooperation for win-win results. China engages in active contacts with other Asian countries on defense and security issues, takes full part in regional and sub-regional security cooperation and acts to promote and build regional multilateral security frameworks. It will seek to provide more security public goods for Asia and the world as a whole.

China is dedicated to resolving hotspot issues in the region. Regarding the DPRK nuclear issue, China has been promoting peace discussions actively and making unremitting efforts to maintain peace and stability on the Korean Peninsula. Interactions among relevant parties on the reopening of the Six-Party Talks are increasing and the situation on the Peninsula is moving towards relaxation. China supports Afghanistan in promoting extensive and inclusive national reconciliation and participates actively in its rebuilding and regional cooperation. China will hold the Fourth Foreign Ministerial Conference on the Istanbul Process of the Afghanistan issue, seeking to make greater contribution to peace and development of Afghanistan and South Asia.

China will continue to properly handle disputes with relevant countries over territorial sovereignty and maritime rights and interests. On issues regarding its territorial sovereignty and maritime rights and interests, China will uphold its principles and bottom line. Bearing in mind the interests of all and with the greatest sincerity and patience, China has always insisted on resolving disputes and divergences through friendly consultations. With respect to the Diaoyu Islands issue, China was forced to take measures to safeguard its sovereignty on solid legal grounds. Establishing the East China Sea Air Defense Identification Zone (ADIZ) is the legitimate right of China as a sovereign state and conforms to international laws and practices.

Along with ASEAN countries, China will continue to implement the Declaration on the Conduct of Parties in the South China Sea (DOC) while promoting the formulation of the Code of Conduct in the South China Sea (COC) in an active and steady manner. We will also implement the important consensuses reached with Brunei and Vietnam, make full use of the China-ASEAN Maritime Cooperation Fund, expand maritime cooperation with relevant countries in an all-round manner, play down our divergences through cooperation and create conditions for the ultimate resolution of the disputes.

(5) Enriching the content of the community of shared future through people-to-people and cultural exchanges. The key to close state-to-state relations lies in the affinity of their people. The people-to-people and cultural exchanges among China and its Asian neighbors have been unprecedentedly active in recent years, with the areas of cooperation expanding and exchanges increasing consistently.

The principle of amity, sincerity, mutual benefit and inclusiveness proposed at the Conference on the Diplomatic Work with Neighboring Countries represents the way China treats its neighbors. "Amity" means to consolidate the friendship that grows from geographical proximity and affection among the peoples. "Sincerity" means to treat people with sincerity and win over others in trustworthy ways. "Mutual benefit" means to follow a vision of cooperation aiming at benefiting the neighboring regions and pursuing mutual benefits and win-win results. "Inclusiveness" refers to a posture of the major countries featuring openness, inclusiveness and seeking common ground while putting aside differences. We need to treat our neighbors as equals and with sincerity, engage in frequent exchanges with them and take heart-warming and heart-winning measures so that our neighbors will be friendlier to us, and recognize and support us more.

China attaches importance to expanding people-to-people and cultural exchanges with its Asian neighbors. By the end of 2012, China has set up 66 Confucius Institutes and 32 Confucius Classes in Asian countries. The amount of exchange students among China and its Asian neighbors reaches almost half a million. The personnel exchanges between China and other Asian countries have exceeded 30 million people. Over 15 million Asian visitors came to China, making up 57 percent of the inbound foreigners to China. Among the top ten sources of foreign visitors to China, seven are Asian countries.

China will continue to expand people-to-people and cultural

exchanges in an all-round way, strengthen exchanges with its neighbors in fields such as culture, tourism, science and education and local cooperation to make more friends and forge new friendships. The year 2014 is the year of Friendly Exchange between China and ASEAN, Mongolia, and India. China, India and Myanmar will commemorate together the 60th anniversary of the Five Principles of Peaceful Coexistence. Working together to make these events successful will enhance mutual understanding and friendship among the peoples.

Catering to the practical needs of Asian countries, China will continue to provide scholarships and conduct training programs, help to train more professionals for other Asian countries. We are ready to exchange experience with other Asian countries on nation building and governance, make the voice of China heard and tell the story of China well. We will strive to link the China dream with that of the people in our neighborhood for a better life and our region for bright development prospects, ushering in a new chapter of people-to-people and cultural exchanges.

V Conclusion

As a new phenomenon in the globalized age, building the community of shared future reflects the general trend of international relations in the new era. Building the Asian community of shared future is the common task bestowed upon Asian nations by history, which will be a reference and an example for the building of the community of shared future for mankind. We need to have wisdom and courage, overcome zero-sum game and Cold War mentality, and open a new path for humanity to make concerted efforts to tackle challenges and realize sustainable development.

By advocating the building of the Asian community of shared future, China carries forward and innovates its good neighborliness foreign policy, and reflects the essence and direction of the major-country diplomacy with Chinese characteristics. China will continue to deepen mutual trust and mutual assistance with other Asian countries, and promote all-round good neighborliness and friendly cooperation. China also hopes that other Asian countries will join in and explore the Asian path, display the Asian wisdom and jointly build a prosperous Asia to make greater contributions to peace and development of humanity.

New Approaches Needed for Tackling Security Challenges in Asia

Remarks at the ASEAN Regional Forum Senior Officials' Meeting
June 12, 2014

Recently, views on Asian security have been permeated with a somewhat gloomy outlook. Some express concerns about growing dichotomy between heated economic exchanges and cool political ties. Others worry about the "trust deficit" in the region. I have just been to the Senior Officials' Meetings for East Asian Cooperation in Myanmar, ASEAN's rotating president, and explored with colleagues possible solutions for security problems in Asia.

Asia has on the whole maintained general peace and stability in the past 20 years since the end of the Cold War. Asian countries have created one economic miracle after another, and made this region the strongest driver of global growth. The ASEAN-led regional cooperation has played an important part in enhancing trust and interaction among major countries in the region. No one should discount the positive mainstream of Asia today.

In the meantime, Asia is facing growing security challenges. The missing Malaysia Airlines MH370, the sunken ferry in the ROK, the Prism scandal last year and the recent terror attack in Karachi Airport are examples of how non-traditional security challenges are directly impacting the lives of ordinary people.

This region also faces some prominent traditional security challenges such as historical grievances, and disputes over territorial and maritime rights and interests. In the age of globalization, outdated security concepts still exist, and certain countries are still clinging to the old security ideas based on military alliances and power politics. The "rebalance" to Asia, which is based on strengthening of bilateral military alliances, has been accompanied by creating new frictions and hostilities in the region. Should Asia fall into division, no one will emerge as a winner. There will only be losers, and regional countries will bear the brunt.

Military alliances often draw a line between allies and non-allies without due regard to the merits of the matter. Support will be given to whatever the ally does, regardless of whether it is right or wrong. As a result, some members of the alliance, with the backing of a strong ally, tend to take adventurous actions by making provocations on maritime issues.

It is explicitly stipulated in the Declaration of Conduct of Parties in the South China Sea (DOC) signed among China and ASEAN members in 2002 that disputes should be resolved by peaceful means among parties directly concerned through consultations and negotiations, and that relevant parties should refrain from taking actions that may complicate or escalate the situation. We hope that the DOC shall be fully and effectively implemented by all its signatories. Pulling extra-territorial countries over to one's side in the disputes will only be counterproductive.

A neighbor of China has chosen to unilaterally submit disputes for international arbitration in violation of the DOC and encroachment on the legitimate rights of China under international law, including the 1982 United Nations Convention on the Law of the Sea (UNCLOS). And we urge another neighbor to immediately stop all forms of disruptions of the Chinese drilling operation in China's own waters so as to restore tranquility in the South China Sea. We will continue to communicate with this neighbor with a view to properly address the current issue.

In the world today, countries are getting more and more interdependent. China and ASEAN members are becoming a community with a shared future, common interests, and common responsibilities. Accordingly we should establish a new security vision to address security problems of the 21st century.

At the recently held Conference on Interaction and Confidence-Building Measures in Asia (CICA) Summit, President Xi Jinping elaborated on the Asian security concept with common, comprehensive, cooperative and sustainable security at the core, and called for a new security architecture in Asia. Common security means respecting and ensuring the security of each and every country. We cannot have security for just one or a few countries while leaving the rest insecure. Comprehensive security means upholding security in both traditional and non-traditional fields, and enhancing regional security governance in a coordinated way. Cooperative security means promoting the security of both individual countries and the region as a whole through dialogue and cooperation. The people of Asia should play a leading role in solving Asian problems. In the meantime, Asia is inclusive and open to the world, and we welcome all parties to play a positive and constructive role in promoting security and cooperation in Asia. Sustainable security means that we need to focus on both development and security to achieve durable security. We need to promote regional integration and foster common progress in both regional economic and security cooperation.

How do we work together to promote peace and security in Asia?

First, we need to deepen regional economic integration, and strengthen the foundation for greater mutual trust. Economic integration in Asia will grow and prosper, and China stands ready to play a greater role in this process.

Second, we need to strive for positive interactions among major countries. Major countries should view each other's strategic intention in a rational way, abandon the Cold War mentality and cooperate with each other to address challenges. This is the expectation of countries in the region, and the ASEAN-led multilateral frameworks such as the ARF and ADMM+ have a positive role to play therein.

Third, we need to properly handle differences and disputes. On the one hand, there should be direct negotiations and consultations among claimants to settle these disputes peacefully and properly. On the other hand, China and ASEAN countries should work together to uphold peace and stability in the South China Sea. We are committed to working with ASEAN countries to implement the DOC in a full and effective way, and steadily move forward COC consultations. In the meantime, we should conduct maritime cooperation such as maritime economy, environment conservation, disaster relief, and maritime security, so as to enhance trust and reduce misgivings, build consensus, and achieve mutually win-win results.

Asian countries are wise and capable enough to solve their own problems and the Asian security concept offers a right perspective.

Fostering a Peaceful and Stable Regional Environment for the Building of the Asian Community of Shared Future

Remarks at the welcoming dinner of the Ministry of Defense for the Fifth Xiangshan Forum on November 21, 2014
Published on the *International Studies*, first edition, 2015

Asia was the cradle of ancient civilizations. Yet the region suffered a lot in modern times, with many Asian countries falling prey to colonial or semi-colonial rule. Many of them won back their independence after WWII, only to be trapped in the Cold War. After the Cold War ended, Asian countries have been actively exploring a path of development suited to their national conditions and realities, following principles of mutual respect, equality and win-win cooperation, and have made solid efforts to promote economic integration. As a result, Asia has made the world-famous miracle and become the world's most dynamic region with the greatest potential. Asian rejuvenation has become the strongest voice of the times.

The in-depth development of globalization has transformed our world into a global village. The interests of countries have never been so closely interconnected. We are all bound together as one community of shared destiny. While we take pride in Asia becoming a major pillar and driver of growth for the world economy, we are aware of the imbalances building up in Asia, i.e., imbalances in Asia's development, between the two wheels of economy and security, and the interests of regional countries and outside players. Standing at a new historical starting point, Asia can only go far if it follows the trend of our times, seize the opportunities and build a closer community of shared future.

In the past two years since taking office, the new Chinese leadership has placed much priority on enhancing regional and international cooperation as it pressed ahead with the agenda of comprehensive deepening of reform and opening-up. Chinese leaders put forward a wide range of initiatives and proposals for regional cooperation such as the

Silk Road Economic Belt, 21st Century Maritime Silk Road, the Asian Security Concept, and a community of shared future between China and ASEAN and Asia in general. As a big Asian nation with 14 neighbors on land and eight along the sea, China knows only too well that it depends on Asia for peace and development, and that its hope and future lie in building the Asian community of shared future.

In my opinion, we need to build three pillars for the Asian community of shared future.

First, a community of shared interests, which is the material foundation for the community of shared future. Asian countries should enhance regional cooperation, turn their economic complementarities into strengths for each other's development, and deepen the convergence of interests to achieve common development. Economic integration is the fundamental channel to build a community of shared interests. To this end, the major countries should support ASEAN community-building, make good use of the ASEAN plus one (10+1) and ASEAN plus three (10+3) frameworks to deepen practical cooperation and gradually shape an East Asian economic community.

Second, a community of shared responsibility, which is the security safeguard for the community of shared future. Stability is a blessing while unrest will bring misfortune. Compared with some other regions, Asia is blessed with overall stability. There exists no major threat to lasting peace in Asia, and this situation should be cherished. Asian countries should take up the primary responsibility for the security of their region, enhancing mutual understanding and trust through closer dialogue and cooperation, providing more public security goods, and jointly promoting regional peace and stability.

Third, a community of culture and people. If economy and security are the two wheels driving the Asian community of shared future, then cultural and people-to-people exchange is the spoke connecting the two wheels. Asia is rich in diversity and cultural heritage. We should promote exchange, dialogue, and encourage inclusiveness and mutual learning. We should work hard to increase the flow of the peoples, deepen friendship and turn Asia into a big harmonious family.

Stability in Asia, a major engine for global economic growth, has wider global significance and is drawing global attention. How to foster a peaceful and stable regional environment for the building of the Asian community of shared future deserves our attention.

First, a new security concept needs to be established. Ideas guide actions, and it requires new ideas to address new problems. Facing the ever growing momentum of regional economic cooperation, old security concepts based on the Cold War mentality, the zero-sum game mentality, and worship of force are no longer relevant. Strengthening bilateral military alliances and ensuring absolute security for oneself will only create divisions and confrontations. In the absence of a generally accepted common security concept in Asia, countries should progress with the times, embrace openness and inclusiveness, and explore a new path for Asian security with new ideas and methods.

At the CICA Conference (Conference on Interaction and Confidence Building Measures in Asia) in Shanghai in 2014, President Xi Jinping proposed an Asian security concept based on common, comprehensive, cooperative and sustainable security. Common security means respecting and ensuring the security of each and every country. Comprehensive security means upholding security in both traditional and non-traditional fields. Cooperative security means promoting the security of both individual countries and the region as a whole through dialogue and cooperation. Sustainable security calls for both development and security to make security durable. This security concept is developed from the common security and cooperative security approach long existing in our region. It opens up brand new prospects for security cooperation in Asia. China is both a proponent and practitioner of this Asian security concept. We are ready to work together with regional countries to achieve security for all through win-win cooperation.

Second, there needs to be stable bilateral relations. One cannot choose its neighbors, and it is natural that neighbors encounter problems with each other. All parties should uphold the Five Principles of Peaceful Coexistence, act in the larger interests of regional peace and stability, respect and accommodate each other's interests and concerns, and resolve disputes through peaceful means to ensure friendly coexistence for generations to come. China has signed treaties of good-neighborliness, friendship and cooperation with a number of countries including Russia, Tajikistan, Kyrgyzstan, Pakistan and Afghanistan and is in discussions with ASEAN on a similar treaty. What these treaties achieve is to reaffirm the commitment of countries to good-neighborly relations through the force of law to institutionalize stable bilateral relations.

Territorial and maritime disputes bear on countries' fundamental interests and national feelings, and hence are highly difficult to resolve. The countries concerned should respect international law and historical facts, take into consideration both factors in reality and future development, and refrain from being biased on any side. Disputes that do not have quick solutions may be shelved and parties can go for common development and gradually narrow differences through dialogue and cooperation. This would be a good way to create conditions for future resolution.

Opting for negotiated settlement by parties directly concerned is a valuable experience gained by China through handling boundary and maritime issues in the past six decades. China has settled boundary issues with 12 out of 14 land neighbors. China and India maintain close communication through the Special Representatives on the Boundary Question. Discussions on the boundary issue between China and Bhutan are going smoothly. China and Vietnam have delineated the maritime boundary in the Beibu Bay and are discussing delimitation and common development in the area beyond the mouth of the Beibu Bay. China and the ROK have agreed to officially launch negotiations on maritime delimitation in 2015.

China and ASEAN countries have maintained effective dialogue on the South China Sea issue and reached extensive consensus. We follow the "dual-track" approach, namely, disputes are to be resolved through negotiations by countries directly concerned while peace and stability in the South China Sea is maintained jointly by China and ASEAN countries. This is a realistic and effective way to properly handle the South China Sea issue. Parties are all committed to fully and effectively implementing the *DOC* and concluding a *COC* based on consensus at an early date. "Early harvest" for the *COC* is being discussed. These include the first document on commonalities, and joint maritime search and rescue hotline and senior diplomatic officials' hotline for maritime emergencies. China is committed to deepening trust and cooperation and safeguarding peace and stability in the South China Sea with ASEAN countries.

China and Japan are close neighbors, and the world's second and third largest economies respectively. For reasons clear to all, this relationship encountered serious setbacks in the past two years. The two sides have issued a four-point principle agreement recently. We hope Japan will move in the same direction as China to gradually bring the relationship back on track. China is committed to managing and resolving the Diaoyu Islands dispute through dialogue and consultation. We urge Japan to face up to history and its responsibilities, speak and act with caution, and make concrete efforts to improve relations with its neighbors.

Third, there should be sound interaction between major countries. Asia is where the interests of the major countries are most concentrated and is a region where the forces interact most frequently. Based on lessons from history, we do not favor outside interference in internal affairs of Asian countries. We believe issues among Asian countries should be addressed by countries concerned. Outside intervention would only complicate matters further. At the same time, we believe in Asia being open. We welcome a constructive role by other major countries in regional security. We are ready to work with them to build a stable strategic framework for regional peace and stability.

China-Russia comprehensive strategic partnership of coordination has developed at a high level with cooperation across the board continuously deepening. China-Russia cooperation is based on the principles of equality, mutual trust and support, common prosperity and ever-lasting friendship. This relationship is neither alignment nor targeted at any third parties. It sets a good example for major countries to deepen trust and cooperation in the new era.

China and the United States are both committed to building a new model of major-country relations, which is also the general hope of regional countries. In June last year, President Xi Jinping and President Obama met informally at Sunnylands, during which they reached important agreement on building the new model based on no conflict, no confrontation, mutual respect and win-win cooperation. Last week when President Obama came to China for the APEC Economic Leaders' Meeting and a state visit, the two presidents met again in Beijing and reached important agreement on taking forward substantially the new model through deepening practical cooperation and managing differences. The two sides agreed, in particular, to develop a pattern of positive interaction and cooperation in the Asia-Pacific on the basis of mutual respect. We will follow through on these agreements to enrich the new model and add positive energy to peace and stability in Asia and beyond.

Fourth, there should be proper handling of hotspot issues. Understanding and trust come from dialogue and communication. Stepping up dialogue and coordination is the right way to handle regional hotspots. China will continue to play the role of a responsible major country in this regard.

The Korean nuclear issue bears on peace and stability on the Peninsula and in Northeast Asia. The problem must be tackled both in terms of symptoms and root causes. Efforts should be made to increase mutual trust and address the concerns of parties in a balanced manner. The parties should demonstrate sincerity and be flexible to restart the Six-Party Talks to find a practical, effective solution acceptable to all. China stands for the denuclearization of the Peninsula. We are all for peace and stability on the Peninsula and peaceful resolution through dialogue and consultation. We will continue to make efforts in this direction.

Afghanistan is at a crucial stage of political, security and economic transitions. At the end of October, China hosted the Fourth Foreign Ministerial Conference of the Istanbul Process of the Afghanistan issue, which was the first international conference on Afghanistan since the Afghan elections. The meeting issued the Beijing Declaration and proposed a number of cooperation projects urgently needed on the ground. China will work with all sides to translate the common understanding and outcomes into concrete actions to support peace, stability and reconstruction in Afghanistan at an early date.

Fifth, there should be adherence to the spirit of international law. All countries, no matter big or small, strong or weak, should uphold international law. China is both a faithful follower and active contributor to international law and regional rules and norms. The People's Republic of China was born after WWII, but the new Chinese government inherited and accepted the international legal order based on the UN Charter. China, India and Myanmar initiated the Five Principles of Peaceful Coexistence to reflect the spirit of law in international relations.

China is committed to upholding regional maritime security and order, and works with ASEAN countries to advance COC consultations for its early conclusion on the basis of consensus. Earlier this year, we presided over the adoption of the updated Code for Unplanned Encounters at Sea (CUES) at the Western Pacific Naval Symposium. We signed MOUs with the US on Notification of Major Military Activities

Confidence-Building Measures Mechanism and regarding the Rules of Behavior for Safety of Air and Maritime Encounters. In addition, China has been actively engaged in rules-making in new areas such as cyberspace and outer space, to contribute to forming fair and equitable international rules.

Next year marks the 70th anniversary of the end of World War II. Countries should work together to uphold the outcomes of World War II and the post-War international order. It would be a good occasion to reaffirm our shared commitment to peace and stability through international rule of law. History allows no distortion. What is at stake is the overall trend of regional peace, stability and development. Only by facing up to history and learning historical lessons can a country win trust from its neighbors.

Sixth, there should be efforts to foster regional security architecture. In doing so, we should be fully aware of the diversity of Asia, and not simply copy the model of other regions. It will only work if it suits regional conditions and serves all parties' needs.

Over the past 13 years since its inception, the Shanghai Cooperation Organization (SCO) has played an increasingly important role in upholding regional security and stability. It has effectively coordinated efforts to fight terrorism, extremism and separatism among its members. Although it is yet to fully resolve the Korean nuclear issue, the Six-Party Talks provides a sound framework for addressing this issue. ASEAN-led multilateral security frameworks such as the ASEAN Regional Forum (ARF) and the ASEAN Defense Ministers' Meeting Plus (ADMM-Plus) have gone a long way to deepening regional security dialogue and cooperation. These mechanisms have different strengths and focuses. They should continue to develop alongside each other. At the same time, there should be more links and interaction among them to strengthen coordination.

China will continue to take an active part in multilateral security dialogue and cooperation and provide more public goods for Asia and the world. China supports closer non-traditional security cooperation in the region. In recent years China has become a major driver for cooperation in the ARF, initiating roughly one third of all new projects. China is committed to enhancing regional disaster relief cooperation. We signed an MOU on Disaster Management Cooperation with ASEAN. China and Malaysia will co-host the fourth ARF disaster relief exercises next year. China will work together with countries in the region to uphold the safety and security of sea lanes. Since 2008, 18 Chinese escort missions have been sent to the Gulf of Aden and half of the ships escorted flew flags of other countries.

Building an Asian community of shared future is a worthy endeavor. China is ready to join its neighbors to embark on the journey towards a better future for our region.

Neighborhood Diplomacy: Innovating both Theory and Practice

Published on the *Global Times* in January 2015

In the just concluded year 2014, China actively explored the major country diplomacy with Chinese characteristics, promoted the establishment of a new model of international relations with win-win cooperation at the core, and assumed the responsibility of a major country for promoting human development and progress in the era of globalization. As an Asian country, China closely links its own development with the common interests of its neighboring countries, and integrates the Chinese dream with the Asian dream in the building of a neighboring community with a shared future. Neighborhood diplomacy epitomizes the major country diplomacy with Chinese characteristics and represents theoretical and practical innovations never seen before.

Theoretical innovation focusing on the building of a neighboring community with a shared future

Asian countries, creators of glorious ancient civilizations that led the world, were subject to the bullying of imperialist powers in modern times and were at the frontier of the Cold War confrontation between the two camps led by the United States and the Soviet Union. Therefore, achieving peace and prosperity and realizing overall rejuvenation are the shared objectives of the Asian people, as they embody Asian countries' joint efforts in the historical journey, and represent the future of Asia's development. Standing at the forefront of the times, China has both the traditions and the future of Asia in mind. After proposing the establishment of the China-ASEAN community with a shared future in 2013, General Secretary Xi Jinping stressed at the Central Conference on Work Relating to Foreign Affairs that we should pay special attention to neighborhood diplomacy and build a neighboring community with a shared future. It is an important part of the major country diplomacy with Chinese characteristics; it combines the political wisdom of Asian countries with the ethos of our times; and it points out the direction for the neighborhood diplomacy in a new era. China has also put forward a series of new concepts, such as the way of getting along with neighboring countries, the "riding along" theory and the Asian security concept, to explore the development paths of China and Asian countries from the political, economic and security perspectives.

The way for neighboring countries to get along is to respect and trust each other, enlarge common ground, resolve differences, and conduct win-win cooperation. All countries should treat each other as equals, respect each other's independence, sovereignty and territorial integrity, respect each other's choice of social systems and development paths, and accommodate each other's major concerns. We should build consensus through dialogue and cooperation, resolve differences and provide the most effective guarantee for long-term stability of the region. We should pursue strength through unity and mutual help, expand the convergence of interests, and achieve common development.

The "riding along" theory means that we welcome Asian countries on board the high-speed train of China's development. China is the world's second largest economy, and the largest trading partner and investor of most Asian countries. China's development is part of the overall development of Asia and provides opportunities for the world, especially its Asian neighbors. Some countries worry that China only takes advantage of other countries' development without shouldering its responsibilities. These worries are really unnecessary. China pursues the greater good and shared interests, seeks development without shirking its responsibilities, and combines justice with interests while giving priority to justice. China relies on the neighboring areas for important support to its development, and is willing to benefit the neighboring areas with its development dividends.

The Asian security concept advocates common, comprehensive, cooperative and sustainable security. China is ready to work with other Asian countries to respect and protect the security of every country, maintain security in both traditional and non-traditional fields in a coordinated manner, promote security of all countries and that of the whole region through dialogue and cooperation, and attach equal importance to development and security in order to achieve lasting peace. The Asian security concept is based on win-win cooperation rather than zero-sum game, and sharing rather than confrontation. It aims at exploring a historic new way to help countries get out of security dilemma and seek long-term stability through joint efforts.

Making the neighboring areas the primary platform for building major country diplomacy with Chinese characteristics

Great acts are made up of small deeds, and every cause of the past and the present can only be accomplished with concrete efforts. Guided by the goal of building a neighboring community with a shared future, the grand layout of neighborhood diplomacy has contributed to the development of the major country diplomacy with Chinese characteristics.

Exchanges with neighboring countries have been close and frequent. China and its neighboring countries enjoy frequent high-level exchanges. In 2014, there were more than 110 visits or meetings at or above the deputy-prime-minister level among China and all of the 23 Asian countries in its neighborhood. All-round cooperation has been carried out at various levels and in various fields, contributing to greater converging interests and growing mutual understanding and trust.

There have been many highlights of neighborhood diplomacy at the multilateral events hosted by China. China has invited Asian leaders to China for the CICA Summit, the APEC Economic Leaders' Meeting, the Dialogue on Strengthening Connectivity Partnership, and the Nanjing Youth Olympic Games, thus elevating good-neighborly friendship and mutually beneficial cooperation to a new level. The Conference Marking the 60th Anniversary of the Five Principles of Peaceful Coexistence was held jointly by China, India and Myanmar, which injected new vitality to the Five Principles of Peaceful Coexistence in the new era.

Cooperation initiatives have benefited the neighborhood. China has advanced Belt and Road cooperation, set up the Silk Road Fund, and made good preparations for the Asian Infrastructure Investment Bank. Great progress has been made in building the China-Pakistan Economic Corridor, and development of the Bangladesh-China-India-Myanmar Economic Corridor proceeded steadily. The series of initiatives have brought tangible benefits to the neighborhood and won the hearts and minds of the people.

The role of a responsible major country has been highlighted. China has successfully promoted a soft landing of major hotspots and tough issues in an effort to maintain peace and stability in neighboring areas. We have played a constructive role in regional affairs by staying firmly committed to denuclearizing the Korean Peninsula, maintaining peace and stability on the Peninsula, and resolving the issue through dialogue and consultation, and by holding the Ministerial Conference of the Istanbul Process of the Afghanistan issue. In response to the freshwater shortage crisis in the Maldives, we have offered all-round assistance from air and sea, demonstrating with concrete actions what a reliable partner should do. On the issue of the South China Sea, we have firmly safeguarded territorial sovereignty and maritime rights and interests. At the same time, we have put forward a dual-track approach, emphasizing the need for the parties directly involved to resolve disputes peacefully through consultation while China and ASEAN countries work together to maintain peace and stability in the South China Sea and promote maritime cooperation. This approach has won the understanding and support of most countries in the region.

Regional cooperation has yielded fruitful results. China has actively worked for the launch of negotiations on the Treaty on Good-neighborliness, Friendship and Cooperation between China and ASEAN countries, the first round of negotiations on an upgraded version of the China-ASEAN FTA, and the start of substantive consultation on the Regional Comprehensive Economic Partnership. To further advance cooperation with South Asian countries, China has announced its plan to invest US$30 billion in South Asia in the next five years and provide US$20 billion in concessional loans. China has increased its engagement in such frameworks as the ASEAN Regional Forum and the Asia Cooperation Dialogue to lead regional cooperation.

Positive interactions among major countries have been promoted. Asia is the home of Asians and an important region for global development. China welcomes the constructive role played by major countries both inside and outside the region to achieve win-win results for all parties. China has deepened strategic coordination and steadily enhanced mutual trust and cooperation with Russia. Within the frameworks of developing a new model of China-US major country relationship and the China-Europe comprehensive strategic partnership, China has expanded cooperation with the United States and the European Union on Asian affairs to resolve differences through dialogue and jointly safeguard peaceful development in Asia.

Striving to create a bright future for the great development of neighborhood diplomacy

China is standing at a new historical starting point, and its neighborhood diplomacy has great potentials. From a dialectical and holistic perspective, we should combine our own development with the common development of the neighboring areas, contributing wisdom and strength to neighborhood affairs while developing ourselves; we should combine economic development with security to safeguard development with peace and ensure peace through development; we should combine bilateral cooperation with regional integration to better coordinate bilateral and multilateral cooperation; and we should combine Asia's centrality with its openness to strengthen cooperation with countries outside the region while further developing partnerships with neighboring countries. This is what we call the "Four Combinations".

We need to work with great enterprise, actively promote the major country diplomacy with Chinese characteristics, make explorations and innovations, deepen cooperation and pursue win-win outcomes in the neighborhood, so that the sense of a neighborhood community with a shared future can take root and bear fruit, both the Chinese dream and the Asian dream can be fulfilled, and new contribution can be made to peace and development in Asia and beyond.

Steadily Advancing Regional Security Cooperation

Speech at the annual meeting of China National Committee of the Council for Security Cooperation in the Asia-Pacific (extract)
January 2015

Strengthening regional security cooperation is an important aspect in tackling security challenges around us. Since joining the ASEAN Regional Forum (ARF) in 1997, we have made remarkable achievements in promoting regional security cooperation as we now speak louder and have greater influence in regional security affairs. In the future, efforts should be made in the following aspects:

First, we should actively promote the Asian security concept. President Xi Jinping proposed at the CICA Summit last year the Asian security concept featuring common, comprehensive, cooperative and sustainable security in Asia, which has attracted wide attention. It conforms to the shared aspiration and reflects the consensus of regional countries on addressing security challenges through cooperation. There are close, inherent links with the common and cooperative security upheld by the ASEAN countries. On the other hand, some countries such as the United States have deliberately made one-sided misinterpretations of the Asian security concept, claiming that China seeks to exclude the United States from Asian affairs by stating that "Asia is for Asians", or that this is merely a Chinese version of the "Monroe Doctrine". They have selectively interpreted President Xi's statement, ignoring the fact that he also emphasized that Asia is an open Asia and it welcomes all parties to play a positive and constructive role in regional security and cooperation. This is a point that I reiterated at the Xiangshan Forum this year.

We must be aware that the Asian security concept is in conformity with the trend of the times and the interests of regional countries, and it gives us a moral high ground on the issue. Foreign Minister Wang Yi stated at the ARF Foreign Ministers' Meeting last year that this new approach to regional security has broken new ground for security cooperation in Asia. In the future, we should continue to actively advocate and promote the Asian security concept for wider acceptance. Of course, we should also do more to increase understanding to dispel mistrust of some countries. Additionally, we need to carefully study how to respond to the increasingly diverse regional security risks and challenges under the guidance of the Asian security concept, putting this idea into practice and making it a consensus and widely-accepted norm for regional countries.

Second, we should make efforts to promote the establishment of a comprehensive regional security structure. Building a security structure in line with the regional conditions is crucial to maintaining long-term peace and stability in Asia. Given the diversity of Asia, it is unlikely for the region to build a highly institutionalized security organization or system similar to that of the European Union in a short time. However, a number of security cooperation frameworks have been established in various sub-regions of Asia, including the ASEAN-led ARF and ASEAN Defense Ministers' Meeting-Plus (ADMM Plus), the Shanghai Cooperation Organization (SCO) and the Six-Party Talks on the nuclear issue on the Korean Peninsula. Although these cooperation frameworks seem to be relatively loose and non-binding, they are open and inclusive, and thus can accommodate the concerns and comfort levels of all parties. Upholding the principle of cooperative and common security, these frameworks represent the development direction of regional security cooperation. With their growing influence, they will become an important foundation for any future regional security structure.

At present, the bilateral military alliance system of the United States still dominates security issues in Asia and even the Asia-Pacific region. However, it is an exclusive system, as its purpose is to maintain only the security of countries within the alliance. The concept lags behind the trend of the times and is seeing waning influence. In recent years, the United States has adopted the "rebalancing" strategy in the Asia-Pacific region, strengthening the alliance system, highlighting military factors, and intervening in territorial and maritime disputes. In response to these actions, we have expressed our concerns on various occasions, but we all know that the United States will not easily abandon the alliance system, which is fundamental to its presence in the region. The main point of our criticism of the US military alliance system is that the system should not target any third party or be involved in regional disputes.

The evolution of the regional security structure in the future depends on the interaction between the multilateral security cooperation frameworks and the US military alliance system. The priority of our work should be promoting the integration of the two systems, reducing friction, and avoiding confrontation by all means in order to minimize the negative impact of the US alliance system, so that the two can play a positive role together in the establishment of a security structure that suits the regional conditions.

Third, we should take concrete steps to promote non-traditional security cooperation. We should focus on the specific needs of regional countries and start from specific areas. If security cooperation in specific areas is done well and, over time, can be done in a systematic and mutually reinforcing way, the future regional security structure in Asia will naturally come into shape.

The most realistic and prominent security threats that Asia faces come from non-traditional security areas, including natural disasters, terrorism, transnational crime, cyber security, and other challenges that are increasingly diverse and complicated. As cooperation in the field of non-traditional security is less sensitive, more inclusive and non-confrontational, and serves the common interests of regional countries, it has increasingly become the focus of multilateral security cooperation frameworks in the region. China has actively participated in cooperative actions in disaster relief, counter-terrorism, combating transnational crime and maritime security, and has undertaken a large number of cooperation projects, assuming an increasingly prominent role. An important part of our efforts to promote non-traditional security cooperation is to provide public security goods through international cooperation, which not only safeguards our expanding interests overseas, but also contributes to the maintenance of regional peace and stability.

We participate in a wide range of non-traditional security cooperation areas, the priority of which is disaster relief and maritime security. Among all regions in the world, the Asia-Pacific is hard hit by natural disasters. Thus, strengthening cooperation in disaster reduction and relief meets the common demand of regional countries. In May this year, we will co-host the fourth ARF disaster relief exercise with Malaysia, which will be the flagship project of regional disaster relief cooperation. It is also the first time for China to go abroad to organize a multilateral military and civilian joint disaster relief exercise. We should continue to actively participate in and lead maritime security cooperation, manage and control crises and safeguard maritime passageways through enhanced dialogues and exchanges. Last year, the annual meeting of the Western Pacific Naval Symposium held in China adopted the new version of the Code for Unplanned Encounters at Sea. We reached consensus with ASEAN countries on the early harvest of the consultation on the Code of Conduct in the South China Sea. We also agreed on the establishment of a joint operation hotline for search and rescue at sea and a senior diplomatic hotline for maritime emergencies. All these are the active efforts we have made to achieve the goals above.

Fourth, we should build a platform for security dialogue that is led by us. At present, Western countries still have the greatest say in regional security affairs. We must accelerate the development of a security dialogue platform that has global influence and is led by the regional countries. Last year, the Xiangshan Forum was upgraded to Track 1.5, and attracted officials and scholars from all over the world, making a good start for the Forum. In the future, the MFA will cooperate closely with hosts from the Chinese military to make the Xiangshan Forum larger and greater. At the same time, we should continue to actively participate in the Shangri-La Dialogue, the Asia-Pacific Round Table Forum and other regional multilateral security dialogues, and use these platforms to promote our ideas and proposals.

CSCAP, the most influential think tank network on security issues in the Asia-Pacific region, is an important platform for us to present our policies, increase understanding and dispel doubts, one that we should manage well and make good use of. CSCAP China has taken active part in Track 2 diplomacy and achieved remarkable results, playing a unique role in China's full participation in regional security dialogue and cooperation. In the future, it is hoped that CSCAP China will continue to enhance its capacity-building, improve its research on regional security issues, promote exchanges and cooperation with foreign think tanks and scholars, and communicate our policies to the rest of the world, providing greater support for Track 1 diplomacy. At the same time, we can explore using CSCAP China as the platform to facilitate the formation of a network of specialized think tanks on Asian security issues with the help of the premium resources of the China Institute of International Studies (CIIS) in order to coordinate Track 2 dialogue and cooperation and better contribute to the neighborhood security cooperation. The MFA will continue to support the work of CSCAP China and we hope that the CIIS will give more support to CSCAP China and make this platform even better.

Building an East Asia Partnership of Win-win Cooperation

Remarks at the "Asia-Pacific Geopolitics" Munich Security Conference
February 6, 2015

Over the past 70 years since the end of World War II, one of the leading trends in global geopolitics has been the rising role of the Asia-Pacific region. East Asia became the third pillar of the global economy together with Europe and North America. Talking about the geopolitics in the Asia-Pacific, the basic question is how to assess the regional situation. But the core issue, however, is whether East Asia can keep its prosperity and stability. This is a question drawing attention from many countries.

One needs to take a multidimensional perspective to come to an objective conclusion about the situation in East Asia. In history, this area has experienced both "hot wars" and the Cold War with certain remnants still haunting the area until today. But the general trend is that the East Asia region has been moving from conflict to peace, from confrontation to cooperation and from poverty to prosperity. There is hope that this trend will continue. Looking around the world, the globe is far from being peaceful. East Europe and the Middle East are suffering from war again while the deep-seated impacts of the international financial crisis are still lingering. By contrast, East Asia has maintained overall stability and fairly fast development and become a major engine driving global economic growth. It is fair to say that East Asia is a region of hope with the greatest dynamism for development.

The rise of East Asia is part of a larger story of rejuvenation of Asia. After Japan, the Republic of Korea, China, and Southeast Asian countries have been rising as a group. This collective development is attributive to multiple factors.

First, East Asian countries have actively engaged in economic globalization in the past two decades, drawing on each other's advantages to achieve common development. Such a peaceful way of collective development demonstrates the democratization of international relations, and makes the balance of power more even, thus providing a stronger foundation for peace and stability in the region.

Second, East Asian countries have prioritized development and the improvement of their people's livelihoods. They have found a path of development that suits them well and the "master key" that can tackle various complex challenges.

Third, East Asia has followed the spirit of openness and inclusiveness in making solid efforts to promote regional cooperation and economic integration. It has actively expanded inter-regional cooperation and achieved integrated development. ASEAN will set up a community this year. And vibrant regional cooperation frameworks such as ASEAN+1, ASEAN+China, Japan and the ROK, China-Japan-ROK cooperation and the EAS have taken shape in the region.

In promoting regional cooperation, East Asian countries have fostered the spirit of cooperation that calls for mutual respect, consensus building, and accommodating the comfort level of all parties, and the vision and habit of win-win cooperation. These are valuable assets gained through efforts to maintain friendly relations and properly manage differences. They also provide a way out to sustain the prosperity and stability of the Asia-Pacific region.

Compared with Europe, East Asia is defined by its diversity in political systems, religions, culture, historical traditions and level of development. As we see from the East Asian experience, those who share the same vision and values are partners. Those who seek common ground while shelving differences can also be partners. At the Bandung Conference six decades ago, Chinese Premier Zhou Enlai proposed that Asian nations seek common ground while shelving differences and pursue peaceful coexistence. This principle has served Asia well in the past 60 years.

In my humble view, the key to continued prosperity and stability of East Asia lies in embracing the spirit of cooperation for win-win results.

First, political equality is the basis for win-win cooperation. Countries, no matter big or small, rich or poor, all have the right to choose their own mode of development. No country should impose its own mode on others.

Major countries should perceive each other's strategic intentions in a rational way. China has been working with the US to build a new model of major-country relationship based on no-conflict, no-confrontation, mutual respect and win-win cooperation. Both sides are committed to strengthening cooperation in the Asia-Pacific. The China-Russia comprehensive strategic partnership of coordination has shown a strong momentum. This relationship is based on non-alignment, and not targeting any third party. China is ready to further develop its strategic relationship of mutual benefit with Japan on the basis of the four political documents between the two sides.

Medium and small countries should not take sides, nor should major countries seek to establish spheres of influence. Regional affairs should be handled through equal consultations. They can no longer be dominated by any country or bloc of countries.

Second, economic integration is the foundation for win-win cooperation. ASEAN and China, Japan and the ROK formulated a vision for an East Asia Economic Community by 2020. The RCEP and the China-Japan-Korea FTA negotiations will go a long way to fulfilling this vision. East Asia regional cooperation, as part of Asia-Pacific cooperation, is a good complement to APEC. Asia-Pacific countries should embrace open regionalism and align their development across the Pacific. TPP and RCEP can be pursued in a coordinated way. They both contribute to the building of an Asia-Pacific FTA. Countries should also promote connectivity of infrastructure, policy-making and of people, to weave a strong network of connectivity in the Asia-Pacific region.

China proposed the Silk Road Economic Belt and the 21st century Maritime Silk Road initiatives and setting up of the Asian Infrastructure Investment Bank to promote regional economic integration. These initiatives are open and inclusive, follow international rules and norms, and operate with modern management. They will be implemented through extensive consultation, joint contribution and shared benefit with interested partners. They can go along with and reinforce existing regional frameworks.

Third, common security provides a major safeguard for win-win cooperation. In the 21st century, we should no longer work on the premise of a zero-sum game between the East and the West. Our security is all tied together. Common, comprehensive, cooperative and sustainable security should be pursued. From the so-called Arab Spring to the "Color Revolutions" in Central Asia to the crisis in Ukraine, there should be some in-depth reflection by the countries concerned. Similar occurrences should not happen in East Asia.

This year marks the 70th anniversary of the end of WWII. Historical perception will be a major issue that bears on the future of the Asia-Pacific. History is never to be denied, still less falsified. Only by facing up to history can one look to the future. In a globalized world, peace and development are the only viable ways. Any deviation from this track will bring disastrous consequences. Germany has set a shining example in this regard. What it has done won respect from its neighbors and the world in general. The relevant countries should learn from this example and not let the historical issue be a strategic burden to the Asia-Pacific.

The key to lasting peace and tranquility in the Asia-Pacific lies in the building of a security structure that meets regional realities. In contrast to Europe, it would be difficult for the Asia-Pacific to put in place a highly institutionalized security system covering the whole region in the near future. Europe's experience can be a useful reference, but it should not be transplanted in the Asia-Pacific region.

The Asia-Pacific is home to a range of multilateral security frameworks, such as the ASEAN-led multilateral dialogue platforms, Shanghai Cooperation Organization and Six-Party Talks. These frameworks generally follow the principles of common and cooperative security, and are inclusive. They conform to the trend of the times and the needs of regional countries. Their influence has been growing, and they represent the direction of regional security cooperation.

East Asia faces territorial and maritime disputes, which bear on the fundamental interests and national feelings of relevant countries. The most effective way to address these disputes is consultations or negotiations by countries directly concerned on the basis of respecting international law and historical facts. Outside intervention will only complicate the situation. Countries should strengthen dialogue and effectively manage disputes before they are settled so that small incidents will not affect the bigger region.

In East Asia there are also America's bilateral military alliances. The existence and evolution of bilateral military alliances bear on how East Asian security cooperation and economic integration can develop in coordination. It is our hope that these alliances will advance with the times, and cooperate more with multilateral security frameworks to jointly contribute to building a security architecture that meets the needs of all sides.

The nuclear issue on the Korean Peninsula is vital to peace and stability on the Korean Peninsula and in Northeast Asia. Parties concerned should show sincerity, and take a more flexible and pragmatic approach to restart the Six-Party Talks as early as possible to find a solution acceptable to all. China supports a nuclear-free Peninsula, peace and stability on the Peninsula, and peaceful resolution through dialogue and consultation.

The most real and salient security risk for East Asia actually comes from non-traditional fields. We need to work together to fight natural disasters, terrorism, transnational crimes and challenges from cyber space, and provide more public security goods. Structures can be fostered to reinforce one another to help lay the foundation for security architecture in the region.

As a major country, China is fully aware that its peace and development is closely tied to that of the Asia-Pacific region. On the other hand, the prospect of the Asia-Pacific is also linked with the future of China. We are ready to work with countries in the region to forge an East Asia partnership of win-win cooperation and together build an Asia-Pacific of lasting prosperity and stability.

Uphold Win-win Cooperation and Promote Peace and Stability in the Asia-Pacific

Speech at the welcoming dinner of the Ministry of Defense for the sixth Xiangshan Forum in October 2015
Published on the *International Studies*, sixth edition, 2015

This year marks the 70th anniversary of the victory of the world anti-fascist war and the founding of the United Nations. It is also the 60th anniversary of the Asian-African Conference in Bandung and the year of completion of the building of ASEAN Community. In this context, this year's Xiangshan Forum is of great significance. The ongoing discussions at this forum of a new vision for security cooperation in the Asia-Pacific will help to enhance mutual understanding and trust and build consensus for cooperation.

At the global level, in the past seven decades, the international order and system set up after the WWII, which was centered on the United Nations and based on the purposes and principles of the UN Charter, has provided a strong guarantee for world peace and development. Peace, development and win-win cooperation have become rising trends of our times. Countries are more and more interdependent and are becoming a community of shared future.

In the Asia-Pacific, the past 70 years saw wars both hot and cold and the emergence of this region from turbulence to peace, from confrontation to cooperation, and from poverty to prosperity. Today, the Asia-Pacific maintains overall stability and vibrant development. It has become the region with the greatest potential, an anchor for world peace and stability and a foundation for world development and prosperity.

Looking at East Asia, over the past 48 years, ASEAN has grown ever stronger, and become an important force for regional peace and development. The ASEAN Community will be established, at the end of this year, as the first sub-regional community in Asia's history and a milestone for East Asia cooperation. It will also lay a solid basis for the building of the East Asia Community and an Asian community of common future.

The stability and prosperity in the Asia-Pacific and in the world has not come by easily and must be cherished. It cannot be ignored that, with the fast changing global political and economic landscape, factors of instability have been on the rise. There is still injustice in international relations. The trust deficit remains acute among some countries. Disputes on territorial sovereignty and maritime rights and interests are being played up. New challenges from natural disasters, terrorism, transnational crimes, cyber security, public health, and energy and resource security keep emerging. There is also a question for East Asian countries: what lessons could we draw from the conflicts and turbulences in West Asia and North Africa?

The new situation calls for new thinking. On the new starting point in history, all countries should think seriously about how to handle state-to-state relations, and open a brighter future for global peace and development with new thinking.

At the end of last year, President Xi Jinping of China called for a new type of international relations featuring win-win cooperation. At the 70th session of the UN General Assembly last month, President Xi elaborated on his thinking on this new type of international relations, by highlighting the following five points:

- building a partnership of equality, mutual consultations and mutual understanding
- forging a security pattern of fairness, justice, broad participation and sharing
- seeking open, innovative, inclusive and mutually beneficial development
- promoting harmonious but differentiated and inclusive exchanges among civilizations
- fostering an ecosystem of respecting nature and green development

This has become the so-called "five-in-one" blueprint for building a community of shared future for mankind.

China is the first major country to take win-win cooperation as the objective for handling relations with other countries. This thinking has its roots in China's broad and profound civilization, in decades of China's diplomatic theories and practice, and from China's long-standing commitment to world peace and development. This new type of international relations carries forward the purposes and principles of the UN Charter, and aims towards the vision of building a community of shared future for mankind. It will be a guiding principle for China's conduct of international relations.

East Asian countries have evolved certain principles and modalities of cooperation, including mutual respect, consensus building and accommodating the comfortable level of all parties. They also cultivated a habit of win-win cooperation. We should reaffirm to these principles, make them deeply rooted in the hearts of the people and part of the values that we all abide by.

As an important member of the Asia-Pacific family, China knows well that its peaceful development is closely linked with the future of the Asia-Pacific. We have always worked hard to promote regional stability and prosperity. China is ready to pursue security dialogue and cooperation in the spirit of win-win cooperation, and jointly safeguard peace and stability in the region with other regional countries. In my humble view, countries in the region should make joint efforts in the following aspects:

First, we should establish a new type of security concept for Asia-Pacific peace and stability. In the 21st century, countries increasingly share common interests, face common challenges and have a common need for better governance. The use of force and confrontation does not get anywhere. It would only make matters worse. Win-win cooperation is the only way forward. We should adhere to the new concept of win-win and all-win, abandon the old thinking of zero-sum game, and pursue a new path of security for the Asia-Pacific, based on joint consultation, broad participation and common benefit.

China put forward the Asian security concept based on common, comprehensive, cooperative and sustainable security. It advocates dialogue and consultation, instead of threat of force. It advocates openness and inclusiveness, instead of exclusion. It advocates win-win cooperation, instead of zero-sum game. This security concept goes with the historical trend of the times and is rooted in regional integration. It gathers the wisdom and consensus among regional countries, reflects the urgent need of all parties to deal with security challenges through cooperation, and opens broad prospects for regional security cooperation.

Second, we shall promote the building of partnerships and strengthen the political foundation for peace and stability in the Asia-Pacific. Asia-Pacific countries are diversified in a unique way. Countries may become partners when they have the same values and ideals, but they can also be partners on the basis of seeking common ground while preserving differences. The key is to adhere to equality and win-win cooperation. Any country, whether big or small, is an equal member of the international community and has the right to choose its development path in line with its national conditions. Regional issues should be handled by regional countries based on equal consultation. Big countries should not seek sphere of influence, while middle and small countries should not take sides among big countries. All parties should make joint efforts to pursue the new path of dialogue instead of confrontation and to pursue partnership rather than alliance, and to build an Asia-Pacific partnership featured on mutual trust, inclusiveness and win-win cooperation.

Positive interaction among major countries is an important part of building regional partnerships. Recently, President Xi Jinping paid a successful state visit to the US at the invitation of President Obama. The two leaders agreed that two countries should continue to put efforts on building the new model of major country relations featuring on mutual respect and win-win cooperation. They also agreed to hold a positive view of each other's strategic intentions and enhance strategic trust, avoid strategic miscalculation and misunderstanding. The visit ushered in a new stage of cross-Pacific cooperation and served as a good example for building the new model of international relations. Based on equality and mutual benefit, China-Russia comprehensive strategic partnership of coordination follows the principle of non-alignment, and not targeting at any third party. Bilateral relations maintain development at a high-level. Committed to building closer partnership for development, China and India strengthened practical cooperation in various areas, and effectively managed the border issues. The strategic and cooperative partnership between the two countries is brimming with new vitality.

For partners, differences and disputes should be handled through friendly consultation in a spirit of mutual understanding and mutual accommodation. As for the disputes with some neighboring countries over territorial sovereignty and maritime rights and interests, China will continue to seek peaceful settlement through dialogue and negotiation on the basis of respecting historical facts and in accordance with international law. Pending the resolution of the disputes, we will work with relevant countries to strengthen crisis management to prevent escalation of disputes. Past progress has shown that enhancing pragmatic cooperation is an effective way to ease differences. This year is the year of China-ASEAN Maritime Cooperation, and our cooperation in this regard has achieved fruitful results. China is ready to work with ASEAN countries in building the 21st Century Maritime Silk Road and promoting common development in the South China Sea to make the South China Sea a sea of peace, friendship and cooperation. China's construction activities and maintenance on some of China's stationed islands and reefs of the Nansha Islands fall entirely within China's sovereignty, and are not targeted at any country. In addition to necessary defense purposes, those facilities are mainly for civilian purposes and serve the common good of the international community. On October 9, China held a launching ceremony of the two lighthouses on Huayang Reef and Chigua Reef of the Nansha Islands. The two lighthouses will provide effective route guidance and navigation aid to vessels passing the surrounding waters, greatly improving the navigation safety in the South China Sea. China will continue to build other civilian facilities serving public interest on its garrisoned islands and reefs of the Nansha Islands to provide better services to passing vessels of the littoral states and all other countries.

Third, we need to promote interconnected development and lay a solid economic basis for peace and stability in the Asia-Pacific region. To expand the convergence of interests is an important basis for sound state-to-state relations. Common and interconnected development provides fundamental safeguards for peace and stability, and holds the master key to various types of security issues. With weak recovery of the world economy, new turbulence on global financial markets and rising downward pressure on the Asia-Pacific economies, it is all the more important for countries to stay focused on development and cooperation.

We need to uphold open regionalism, and make all trade arrangements and cooperation mechanisms in our region complement each other and advance in parallel. East Asia cooperation is an important part of Asia-Pacific cooperation. TPP negotiation has just been concluded. We should work hard to complete the negotiations for the RCEP and the upgrading of China-ASEAN FTA by the end of the year.

There should be better complementarities and coordination between the various regional trade arrangements so that, together, they could forge a synergy in pushing forward the building of an Asia-Pacific free trade area, shaping an open, interconnected, balanced, and win-win economic framework in the Asia-Pacific and injecting new vigor into the global economy. China is taking active steps to build the Silk Road Economic Belt, the 21st Century Maritime Silk Road and the establishment of the Asia Infrastructure Investment Bank. These initiatives are all aimed at promoting interconnected and common development. They are open and inclusive, and will move forward together with the existing multilateral mechanisms and initiatives. We welcome all parties to participate in these initiatives for shared benefits.

Fourth, we need to promote the rule-making and improve the institutional safeguards for peace and stability in the Asia-Pacific region. As a saying goes, nothing can be accomplished without norms. For countries to live together in peace, it is also imperative to observe the spirit of rule of law, abide by the international norms based on the purposes and principles of the UN Charter, and subscribe to those widely recognized and fair rules. International and regional rules should be formulated and observed by all countries concerned, rather than dictated by any particular country. Rules of individual countries should not automatically become "international rules", still less should individual countries be allowed to violate the lawful rights and interests of others under the cover of so-called rule of law.

China is committed to upholding international rules and has been a constructive player in this system. China has joined almost all inter-governmental organizations and acceded to over 400 international and multilateral treaties. China has signed treaties of good-neighborliness, friendship and cooperation with eight neighboring countries. We are now in the process of discussion with ASEAN countries on the signing of a Treaty on Good-Neighborliness, Friendship and Cooperation, and we are ready to sign such bilateral treaties with all ASEAN members when the conditions are ripe.

China is also committed to safeguarding the maritime security and order in the region and enhancing the building of institutions and rules. We will continue to work with ASEAN countries to fully and effectively implement the DOC, and steadily work for early conclusion of a COC on the basis of consensus. At the China-ASEAN Defense Ministers'

Informal Meeting yesterday, China's proposal on holding a joint training with ASEAN countries under the Code for Unplanned Encounters at Sea (CUES) in the South China Sea next year was welcomed by ASEAN members.

Fifth, we need to promote the building of regional security architecture and consolidate institutional support for peace and stability in the Asia-Pacific. Security cooperation in the Asia-Pacific has long being lagged behind to economic cooperation in the region. It is high time now to build an open, stable and widely acceptable regional architecture for security cooperation that fits the reality of this region.

China has long worked to push forward the building of regional security mechanisms. We initiated with relevant partners the Shanghai Cooperation Organization and Six-Party Talks, established the Xiangshan Forum, successfully hosted the Summit of Conference on Interaction and Confidence Building Measures in Asia (CICA) and the Foreign Ministerial Conference of the Istanbul Process on Afghanistan issue, and actively participated in the ASEAN-led multilateral security dialogues and cooperation mechanisms. In my opinion, with each country having its own focus and features, these mechanisms and platforms all subscribe to the outlook of cooperative security and common security. They are inclusive, meet the trend of our times, fit the needs of regional countries, and represent the future direction of regional security cooperation. Together, they will form an important basis for the creation of future regional security architecture. China will continue to support the capacity and institution building of CICA, and encourage all parties to enhance dialogues and exchanges under the East Asia Summit to accumulate consensus and advance the building of the regional security architecture in a step by step approach.

China supports bolstering pragmatic cooperation in non-traditional security areas and giving more substance to regional security architecture by enhancing institution building for cooperation in various areas. In recent years, China has put forward many cooperation initiatives in non-traditional security areas under various regional mechanisms, giving a strong boost to the exchange and cooperation in these areas. At the China-ASEAN Defense Ministers' Informal Meeting yesterday, China proposed to hold joint maritime search and rescue and disaster relief exercises next year to raise the capacity for jointly tackling maritime challenges.

The Chinese people are working hard to realize the Chinese dream of the great renewal of the Chinese nation. This will be a process that brings greater opportunities and benefits for the development and cooperation in China's neighborhood, the Asia-Pacific region and the world at large. China's development adds to the force for world peace. China will pursue the path of peaceful development, stick to the policy of fostering amicable ties with neighboring countries, and work to maintain and promote stability and prosperity in the Asia-Pacific region. We are ready to work with regional countries to pursue win-win cooperation, build the new type of international relations and jointly usher in a brighter future for the Asia-Pacific region.

Work Together for a Future of Peace and Development in Northeast Asia

Remarks at the Northeast Asia Peace and Cooperation Initiative Forum
Seoul, October 27, 2015

The international landscape has undergone profound and complex changes. In this increasingly multi-polar, economically globalized, digitized and culturally diversified world, countries have become more interconnected and interdependent than ever before, moving toward a community with shared interests, shared future and shared responsibilities. On the other hand, however, the world is far from being tranquil with incessant local conflicts and hotspot issues. Countries around the world have found their paths to development bumpy and challenging. The world economy, amid deep-going readjustments, has experienced difficult recovery, coupled with sluggish growth in global trade and investment and volatility in international financial markets.

Last year, the ROK successfully hosted the first NAPCI Forum, which serves as a good platform for Northeast Asian countries to enhance mutual trust and cooperation. Today, representatives are gathered once again in Seoul to discuss ways to advance peace and cooperation in Northeast Asia and promote the common development in the region. I am sure that at this forum, we will be able to build on what has been achieved and have constructive discussions on building mutual trust and working for peace and prosperity in Northeast Asia.

Northeast Asia, a vast land with rich natural endowments, is our shared homeland where we survive and thrive together. The six countries in the region, with a combined population of 1.7 billion and rich land and marine resources, account for one-fifths of the world economy. This part of the world, lying at the crossroads of Eastern and Western civilizations, has a time-honored history spanning thousands of years and is known for its unique wisdom and charm. Various types of economies, developed, newly industrialized and developing ones, are all thriving in the region. The GDP of three regional countries ranks among the world top 11. Northeast Asia, with marked advantages of late-starters, is the most dynamic and promising region in the world and an important engine for world economic recovery and growth.

That said, however, the region is also faced with various difficulties and challenges. Given the differences in political system and development path, countries in the region vary in development level. Issues over history, territory, oceans and resources are found in state-to-state relations. And among certain countries, a lack of mutual trust has led to trust deficit. With the shadow of the Cold War still lingering on, the region sees rising non-traditional security threats such as terrorism, transnational crime, cyber security, climate change, and natural disasters. Regional economic integration process, behind that of EU and NAFTA, has yet to keep pace with the trend of economic and regional integration or to meet the needs of development of regional countries.

Peace and development in Northeast Asia are our common aspiration and dream that call for the joint efforts of all regional countries. As an important member of the Northeast Asian family, China sees its development inseparable from that of the region, and regional prosperity, in turn, needs the input of China as well. China will stay committed to the path of peaceful development and the win-win strategy of opening-up. We will work for a new type of international relations featuring win-win cooperation and remain a staunch force for world peace and common development. China will continue to build friendship and partnership with our neighbors, foster an amicable, secure and prosperous neighborhood, and uphold the principle of amity, sincerity, mutual benefit and inclusiveness, with a view to continuously deepening mutually beneficial cooperation with Northeast Asian countries, and working together for peace and prosperity in the region.

It is important for Northeast Asian countries to jointly build a solid foundation for prosperity and stability in the region. To this end, I would like to make the following proposals:

First, enhancing political and security strategic coordination based on mutual respect and mutual trust. Mutual respect is the precondition for political exchanges. We in Northeast Asia need to respect each other's core interests and major concerns, view other countries' development and policies in an objective and rational way, and strive to seek common ground while shelving or even transcending differences. And on that basis, we need to deepen high-level exchanges and political communication with good faith and sincerity to enhance consensus, and uphold and advance regional peace and stability. Mutual trust underpins security cooperation. We in Northeast Asia need to strengthen strategic coordination, reduce factors that may affect mutual trust, and promote security in the region through dialogue and cooperation. It is important to promote peace and security through cooperation, and resolve disputes peacefully through dialogue and consultation. Equal importance should be placed on development and security. Efforts should be made to promote sustainable security through sustainable development, and realize sustainable development through sustainable security. It is hoped that all parties will work together to truly discard the Cold War mentality and nurture new security concepts as we explore a path for Asia that ensures security for all, by all and of all.

Second, deepening region-wide practical cooperation in economy and trade on the basis of mutual benefit. We in Northeast Asia need to reject the obsolete notion of zero-sum game, and follow a new approach of win-win and all-win cooperation. It is important to strengthen macroeconomic policy coordination, keep expanding trade and investment cooperation, and enhance the level of trade liberalization and facilitation among countries and in the whole region through free trade agreements. On the other hand, efforts should be made by all sides to accelerate regional cooperation processes, including the China-Japan-ROK cooperation, the Greater Tumen Initiative, Northeast Asian regional cooperation, and East Asian cooperation, and to develop regional free-trade, investment, and financial cooperation systems. In this way, countries in the region will be able to achieve their respective development and facilitate regional economic integration at the same time. In this process, the strengths and comparative advantages of all countries should be fully leveraged for international industrial capacity cooperation. In this regard, we in Northeast Asia enjoy sound conditions. Japan and the ROK lead in high-end manufacturing with advanced technology and equipment. Russia, Mongolia, and the DPRK enjoy unique advantages in resources and have strong demand for infrastructure and equipment. China has a good industrial capacity of making quality and cost-effective mid-end equipment. Industrial capacity cooperation will achieve multiple purposes by bringing the strengths of regional countries into full play, meeting the demands of all countries, and lending impetus to regional economic cooperation for mutual benefit.

Third, bringing regional countries closer to each other through enhanced connectivity. Better infrastructure connectivity is of great significance in breaking bottlenecks to development and bringing regional countries closer to each other. We need to actively promote the construction of cross-border railways and bridges, improve the development of oil and gas pipelines, communications and land ports networks to facilitate the movements of goods and people across the region. An important goal of China's Belt and Road Initiative is to enhance the connectivity of participating countries. In this regard, Northeast Asia is an important region to which the Belt and Road cooperation will extend. The participation of regional countries is essential in realizing such a vision of connectivity. Looking ahead, we see connectivity in infrastructure as the first step in our joint endeavor. Efforts can be made to advance region-wide connectivity by enhancing our policy communication, unimpeded trade, currency circulation and people-to-people exchanges. It is hoped that continued progress will be made towards this goal through our concerted efforts.

Fourth, seeking harmonious development and common prosperity of the region through affinity and mutual assistance. All-round development of Northeast Asia depends not only on closer economic cooperation, but also continued deepening of exchanges between our peoples. As the saying goes, the key to state-to-state relations lies in the affection between their peoples. People-to-people exchanges focus on the people, draw inspirations from the people and bring benefits to the people. We in Northeast Asia, by fully leveraging our geographical proximity, cultural affinity and amity, need to further enhance exchanges between our youths, civil societies, local governments, media and think tanks, and expand exchanges in culture, science, education, and tourism, with a view to promoting dialogue among civilizations and exchanges between development models. In this way, we will be able to consolidate the foundation of public support and build closer bonds between our peoples. Given the history and realities of Northeast Asia, regional countries vary in terms of development level and stage. However, with our interests deeply intertwined and interconnected, our countries are closely linked in our development. In line with the principle of win-win cooperation, we in Northeast Asia, developed and developing countries together, need to join hands and work for common development and a community with a shared future.

It is important for regional countries to further deepen communication and exchanges, build consensus, and strengthen cooperation of mutual benefit to embrace a bright future of peace and development in this region. Peace and cooperation in Northeast Asia is of great significance and will not be possible without the wisdom and input of insightful people of all regional countries.

The 21st Century Maritime Silk Road and ASEAN Countries

Remarks at the opening ceremony of the International Seminar on the 21st Century Maritime Silk Road
December 21, 2015

In the autumn of 2013, President Xi Jinping put forward the initiative of jointly building the Silk Road Economic Belt and the 21st Century Maritime Silk Road. Central Asian and Southeast Asian countries as well as the broader international community have appreciated, supported and actively participated in the initiative. The Belt and Road Initiative pursues win-win cooperation in an open and inclusive spirit and follows the principle of extensive consultation, joint contribution and shared benefit, which is in line with China's neighborhood policy of "amity, sincerity, mutual benefit and inclusiveness". With the support of more than 60 participating countries, the Belt and Road cooperation serves as a new platform for these countries to promote mutually beneficial cooperation, common development and prosperity.

The 21st Century Maritime Silk Road could trace its origin back to the time-honored history of exchanges between the peoples of China and Southeast Asia. More than 2,000 years ago, the fleets of the Han Dynasty (206 BC-220 AD) of ancient China sailed far into the South China Sea, opening the door to maritime exchanges between China and other countries and starting a maritime Silk Road. Wang Dayuan, an outstanding navigator of the Yuan Dynasty (1279-1368), visited the Philippines, Brunei, Kalimantan, Java and other places. Back in China, he wrote a book recording his journeys on the sea, helping his Chinese contemporaries learn more about Southeast Asia. Zheng He, a Chinese navigator of the Ming Dynasty (1368-1644), visited Southeast Asia on all his seven voyages overseas. His fleets did not occupy an inch of foreign land. Rather they brought with them silk, porcelain and tea, upheld maritime peace and tranquility, and sowed the seeds of friendship and mutual assistance. The 21st Century Maritime Silk Road inherits and carries forward the Silk Road spirit of peace and cooperation, openness and inclusiveness, mutual learning and mutual benefit.

China and Southeast Asian countries, enjoying geographical proximity, cultural affinity and converging interests, are good neighbors and good partners. China has entered the decisive stage of building a moderately prosperous society in all respects, and will soon implement the 13th Five-Year Plan by speeding up the shift of growth models and promoting a new type of industrialization, IT application, urbanization and agricultural modernization. China's economy will continue to unlock its great potential.

ASEAN countries are also working hard towards the goal of achieving national development and rejuvenation and making their people more prosperous and happier. China and ASEAN countries share similar goals of development, and much can be accomplished in our cooperation.

Southeast Asia is a high priority in China's neighborhood diplomacy and a key region in the building of the 21st Century Maritime Silk Road. China and ASEAN countries have seen their political mutual trust deepening and practical cooperation expanding. Since the beginning of this year, we have formulated the Plan of Action to Implement the Joint Declaration on China-ASEAN Strategic Partnership for Peace and Prosperity (2016-2020). The first China-ASEAN Defense Ministers' Informal Meeting was held in China. The protocol on upgrading the China-ASEAN Free Trade Area was signed. From January to October, bilateral trade volume reached US$379.2 billion, and accumulated mutual investment topped US$150 billion. A number of connectivity projects, including China-Laos Railway and China-Thailand Railway, have made significant progress. The Asian Infrastructure Investment Bank (AIIB), participated by the ten ASEAN countries, will soon start operation. China and ASEAN see our future closely connected, our interests increasingly converging and the bonds between our peoples closer. China and ASEAN countries have the condition and ability to build the 21st Century Maritime Silk Road into a road leading to smooth development, mutual benefit, mutual learning, cooperation, harmony, peace as well as closer people-to-people ties and profound friendship. To deepen cooperation between China and ASEAN countries in this regard, I would like to make the following proposals.

First, promote connectivity. Connectivity, as a priority in our endeavor, provides a firm basis for advancing the building of the Maritime Silk Road. It is important to strengthen our connectivity in infrastructure, basic industries, economic cooperation zones, financial investment, institution and software, and develop a safe, well-coordinated multi-model transportation network that links China and its neighboring countries, laying a solid foundation for common development.

Second, enhance economic and trade cooperation. We need to seize the opportunity of upgrading China-ASEAN FTA to improve the trade structure, introduce new ways of trade, set up and develop a service trade support system, and promote balanced growth of trade. We need to make more efforts in building a currency stability system, investment and financing system and credit information system in Asia, support the AIIB to start operation at an early date and provide important support for the building of the Maritime Silk Road.

Third, build synergy and promote cooperation. In line with the principle of respecting differences and treating each other as equals, efforts should be made to build stronger synergy within our development strategies. Alignment could be made between the Belt and Road Initiative with Indonesia's Global Maritime Fulcrum Strategy, Malaysia's Economic Transformation Plan, and Vietnam's Two Corridors, One Economic Circle Initiative. Moreover, the Plan of Action to Implement the Joint Declaration on China-ASEAN Strategic Partnership for Peace and Prosperity (2016-2020) could align with ASEAN countries' development plans and ASEAN Post-2015 Development Agenda. Such efforts could facilitate free flow and mutual reinforcement of economic factors both over land and at sea, lending new impetus to the development of the Maritime Silk Road.

Fourth, jointly uphold peace and stability in the South China Sea. We need to stay committed to friendship and cooperation, properly handle disputes, manage and control differences. Efforts should be made to vigorously advance maritime cooperation and common development by making good use of the China-ASEAN Maritime Cooperation Fund. We need to fully and effectively implement the Declaration on the Conduct of Parties in the South China Sea, and reach agreement on a code of conduct in the South China Sea at an early date on the basis of consensus building, so that the 21st Century Maritime Silk Road will develop under the sunshine of peace.

Fifth, strengthen cultural exchanges and cooperation on

environmental protection. By tapping on the rich historical and cultural assets of the Maritime Silk Road, we need to enhance cooperation in education, culture, tourism, health, science and technology. The involvement of both public and private sectors in diverse activities will help enhance friendship, build closer bonds among our peoples and lay the foundation of strong popular support for cooperation between the two sides. Efforts should be made to highlight ecological conservation and promote cooperation on environment and climate change with a view to jointly working for a green Silk Road.

The year 2016, marking the 25th Anniversary of China-ASEAN Dialogue Partnership, will be the first year for the Chinese people to move towards the goal set out in our 13th Five-Year plan and the year of all-round development of the Belt and Road cooperation. As the Chinese saying goes, the flame runs high when everybody adds wood to it. Building the 21st Century Maritime Silk Road is a common cause where China, ASEAN countries and other participating countries pitch in together. In this endeavor, all countries need to make concerted efforts to synergize their development strategies and expand common interests. It calls for confidence, resolve, wisdom and action of all countries.

A Symphony of a Shared Future for China and Asia

Published on the *People's Daily* on December 31, 2015

In the year 2015, China has achieved all-round progress in its neighborhood diplomacy. As an important member of the Asian family, China, with a keen appreciation of the domestic and international situations, has vigorously expanded exchanges and cooperation with other Asian countries on the two major themes of peace and development, and worked for an Asian community with a shared future featuring mutual benefit and common prosperity, sending a resounding message of safeguarding peace and promoting development in Asia. Over the year, a host of important achievements have been made in China's neighborhood diplomacy.

Advocating the vision of an Asian community with a shared future

As globalization deepens and regional integration accelerates, Asian countries have become increasingly interdependent with their interests deeply entwined and their future closely connected. At the Boao Forum for Asia Annual Conference this year, President Xi Jinping gave a comprehensive account of the significance of building an Asian community with a shared future and the ways to achieve this goal, and presented the bright prospect of cooperation and development in Asia.

Over the past year, China has put Asia on top of its diplomatic agenda and worked to gather public support in the region for an Asian community with a shared future. Based on equality and mutual respect and following the Asian spirit of harmonious coexistence, good-neighborliness, consultation and dialogue, respect for diversity of civilizations, solidarity and cooperation, China has intensified high-level exchanges and enhanced political trust with other Asian countries.

President Xi Jinping visited Pakistan, Vietnam and Singapore, attended the Asian-African Summit and the Commemoration of the 60th Anniversary of the Bandung Conference in Indonesia, and interacted with leaders of other Asian countries on the sidelines of multilateral summits. President Xi put forward a host of important proposals on China's relations with neighboring countries, South-South cooperation and Asia-Africa cooperation, which have been enthusiastically responded by Asian leaders. Premier Li Keqiang visited the Republic of Korea, and attended the China-Japan-ROK (CJK) Trilateral Leaders' Meeting, which was resumed after a three-year suspension. He also paid a visit to Malaysia and attended the leaders' meetings on East Asian cooperation, giving a strong boost to China's bilateral relations with relevant countries as well as regional cooperation.

Asian integration has provided a major driving force for the building of the community with a shared future. In 2015, negotiations on upgrading the China-ASEAN FTA were concluded, ushering China-ASEAN cooperation into a new phase. China worked with other parties to restart China-Japan-ROK cooperation, with steady progress made in the CJK FTA negotiations. Important headway has been made in negotiations on the Regional Comprehensive Economic Partnership Agreement (RCEP) as evidenced by the goal set by the relevant parties to complete the RCEP negotiations by 2016 and establish the East Asian Economic Community by 2020. China successfully held the first Lancang-Mekong Cooperation (LMC) Foreign Ministers' Meeting, during which the parties reached a consensus on building LMC into a new platform for sub-regional cooperation.

Taking the initiative to promote Asia's development and prosperity

Asia is a priority in building the Belt and Road Initiative (BRI). At the Boao Forum for Asia Annual Conference, China published the Vision and Actions on Jointly Building Silk Road Economic Belt and 21st-Century Maritime Silk Road, which have been welcomed by the neighboring countries. Under the BRI, China has aligned its development strategy with those of neighboring countries to promote cooperation in such areas as infrastructure and industrial capacity and strengthen economic, trade and investment relations. The pilot operation of the BRI in Asia has produced fruitful results.

Significant progress has been made in infrastructure connectivity. China and Indonesia nailed the Jakarta-Bandung High-speed Railway Project and will officially start the construction next year. The launch ceremony was held for the construction of the China-Laos Railway and China-Thailand Railway. China and Vietnam have stepped up efforts to plan the Laokai-Hanoi-Haiphong standard gauge railway route. Progress has been made in China-India cooperation in the speed-up of the existing railway line, station redevelopment, and feasibility study of the New Delhi-Mumbai high-speed railway project.

Development of economic corridors has gained momentum. The China-Pakistan Economic Corridor has been harnessed to drive practical cooperation with a focus on the Gwadar Port, energy, transport infrastructure and industrial cooperation, and a number of major projects were kicked off. The Quartet Joint Working Group on the Bangladesh-China-India-Myanmar Economic Corridor has made new progress in its work, and the plan for China-Mongolia-Russia Economic Corridor is now under development, with the medium-term cooperation road map signed among the three countries.

Production capacity cooperation has accelerated. China has rolled out production capacity cooperation with neighboring countries in an all-round way, making new headway in various projects, including industrial parks, cross-border economic cooperation zones and port industrial zones. A series of activities on China-ASEAN production capacity cooperation were successfully held, making such cooperation a new driver of overall China-ASEAN cooperation.

New financing mechanisms have been established and started operation. The Asian Infrastructure Investment Bank (AIIB) is up and running. Nineteen Asian countries, including China, have joined the bank, accounting for one third of its 57 prospective founding members. The AIIB, which is of ground-breaking significance, has injected fresh impetus into development financing in Asia and global financial governance. The Silk Road Fund has become fully operational with its first key project slated to support the building of China-Pakistan Economic Corridor.

Undertaking responsibilities to defend peace and stability in Asia

The year 2015 marks the 70th anniversary of the victory of the world anti-fascist war and the founding of the United Nations. Over the past seven decades, Asia has emerged from turbulence, confrontation and poverty, and become a region of peace, cooperation and prosperity.

Committed to defending peace and security and upholding order and justice in Asia, China has made vigorous efforts to promote dialogue on security cooperation and proper settlement of regional hotspot issues, becoming a major force underpinning peace and stability in the region.

China has committed itself to upholding peace and common security. China held a solemn event to commemorate the 70th anniversary of the victory of the Chinese People's War of Resistance against Japanese Aggression and the world anti-fascist war. The attendance of the leaders of ten Asian countries at the event fully testifies to the strong consensus of China and other Asian countries to bear history in mind and uphold historical justice, and their collective will of safeguarding the post-war international system with the United Nations at its core. For the first time, China hosted the China-ASEAN Defense Ministers' Informal Meeting and the China-ASEAN Ministerial Dialogue on Law Enforcement and Security Cooperation, and successfully held the sixth Xiangshan Forum. China conducted joint exercises and training together with neighboring countries, and carried out joint patrol and law enforcement operations on the Mekong River. Through these security exchanges and cooperation, China and neighboring countries have turned the "Asian security concept" into concrete actions, and embarked on a path toward common, comprehensive, cooperative and sustainable security in Asia.

China has played its role as a major country in managing hotspot issues. As a responsible major country, China has played an active role in promoting and coordinating efforts to cool down hotspot issues in the neighborhood and provided more public goods for security in Asia. Seizing the opportunity of the tenth anniversary of the September 19 Joint Statement of the Six-Party Talks, China has reached out to the relevant parties to deepen engagement, keep denuclearization process on track and explore a peace and security mechanism on the Korean Peninsula and in Northeast Asia. China has played a constructive role on the issue of Northern Myanmar, and encouraged the parties concerned to engage in contact and peace talks, stabilize the situation in Northern Myanmar, and properly handle sensitive border issues. China has promoted peace talks among the parties in Afghanistan, and taken an active part in the settlement of the Afghan issue. On the South China Sea issue, China has resolutely safeguarded its territorial sovereignty as well as lawful and legitimate maritime rights and interests, and firmly opposed the erroneous actions taken by certain countries to escalate disputes and create tensions. China has supported and advocated the "dual-track approach" to the South China Sea issue, and worked with relevant countries to push for the full and effective implementation of the Declaration on the Conduct of Parties in the South China Sea, well manage differences, advance maritime cooperation, and safeguard peace, stability and freedom of navigation and over-flight in the South China Sea. In the wake of the earthquake and fuel crisis in Nepal, and the earthquakes in Afghanistan and Pakistan, China provided timely assistance to the three countries. These are the concrete actions China has taken to promote the greater good and shared interests.

Playing China's role to open new grounds in neighborhood diplomacy

The year 2016 will be the year for China to kick off its 13th Five-Year Plan for economic and social development. A host of major initiatives will be implemented with early harvests. In the new year, China will speed up efforts to build the Asian community with a shared future and connect the "Chinese dream" with the "Asian dream". Politically, China will strengthen high-level exchanges with neighboring countries, deepen political mutual trust and work for new progress in building partnerships, so that more countries will become China's good neighbors, good friends and good partners. Economically, China, focusing on the BRI, will highlight the two priorities of connectivity and production capacity cooperation, pursue greater synergies of development strategies, and promote complementary and coordinated progress of regional trade arrangements and cooperation frameworks. In the area of security, China will work for the building of robust institutions underpinning peace and stability in Asia, and play its role and make its contribution on hotspot issues. China will firmly uphold peace and stability on the Korean Peninsula and continue to work for the resumption of the Six-Party Talks. China will actively participate in building peace and support internal reconciliation in Afghanistan. On the South China Sea issue, China will safeguard its legitimate rights and interests, adhere to dialogue and consultation, and ensure smooth navigation and stability in the South China Sea.

With the full launch of its 13th Five-Year Plan in the new year, China will forge ahead toward the realization of its first centenary goal, that is, ushering in a moderately prosperous society in all respects by 2020, which will generate strong momentum for national development. China's neighborhood diplomacy aims to ensure a sound external environment for implementing the 13th Five-Year Plan. This will bring new impetus to development and cooperation and enable China to make greater contribution to peace, stability and prosperity of Asia. There are good reasons to expect an even more beautiful symphony of a shared future for China and Asia.

Promoting East Asian Cooperation and Realizing the Vision of Common Prosperity through the Belt and Road Development

Remarks at the opening session of the Ninth Asian Financial Forum
January 18, 2016

In autumn of 2013, President Xi Jinping proposed the major initiatives on the development of the Silk Road Economic Belt and the 21st Century Maritime Silk Road during his visits to Central Asia and Southeast Asia. Over the past two years, the initiative has been increasingly welcomed by the international community, and more than 70 countries and international organizations have participated in the Belt and Road development, sending a most powerful message of peace, development and win-win cooperation in this era.

The Belt and Road is the best manifestation of the win-win spirit. Guided by the principle of consultation, cooperation and benefit for all, the initiative seeks the full blossom not of a single flower, but of all flowers. It is not a Chinese solo, but a chorus of all participating countries. Following the Silk Road spirit of peace and friendship, openness and inclusiveness, mutual learning and mutual benefit, it is part and parcel of China's efforts to build a community with a shared future for mankind, and echoes China's neighborhood policy of amity, sincerity, mutual benefit and inclusiveness. The initiative's five key elements, policy, infrastructure, trade, financial and people-to-people connectivity, form a complete system of cooperation underpinned by three important pillars of software, hardware development and personnel exchanges.

The Belt and Road is a strategic platform for mutually beneficial cooperation among participating countries. In 2015, the Belt and Road set out its vision and started implementation. China has signed Belt and Road cooperation agreements with more than 20 countries. As two major countries on the Eurasian continent, China and Russia have further coordinated their development strategies through a joint statement on cooperation between the Silk Road Economic Belt and the Eurasian Economic Union, established a coordination mechanism, and decided to use the Shanghai Cooperation Organization as the main platform for advancing this goal. China and Europe have agreed to form synergy between the Belt and Road Initiative and the Investment Plan for Europe, and set up a China-EU Joint Investment Fund. China has signed industrial capacity cooperation agreements with countries in Asia, Africa, Latin America and Europe, established bilateral or multilateral cooperation funds, and put in place an extensive network of global industrial capacity cooperation.

The Belt and Road is a powerful engine for the new round of development in China's neighborhood. China's neighboring countries are the main participants and primary beneficiaries in Belt and Road cooperation. The rapid advance of cooperation in neighboring countries has played an exemplary role in Belt and Road development. In Northeast Asia, China and the ROK have promoted cooperation on four development strategies, China and Mongolia have agreed to synergize the Belt and Road and the Steppe Road, and China, Russia and Mongolia have formulated a medium-term roadmap for the development of an economic corridor connecting the three countries. In Southeast Asia, construction has started on industrial parks, cross-border economic cooperation zones and port-vicinity industrial parks, and the China-Indonesia Jakarta-Bandung High Speed Railway, China-Laos Railway and China-Thailand Railway have been launched one after another. In South Asia, China and India have strengthened cooperation in infrastructure and other areas, a large number of key projects have started on the China-Pakistan Economic Corridor, and progress is being made on the Bangladesh-China-India-Myanmar Economic Corridor. As an important financing mechanism for the Belt and Road, the Silk Road Fund has made successful investment on substantive projects. The inauguration ceremony of the Asian Infrastructure Investment Bank (AIIB) was held two days ago, and 57 countries became its founding members.

A good beginning is half the task. The early harvest achieved in the Belt and Road development has laid a solid foundation for its full advancement in the future, injected new dynamism to the development of China's neighborhood and the whole Eurasian continent, and added new impetus to global economic recovery.

148

The Belt and Road is an inclusive and open system that is complementary and mutually reinforcing with existing cooperation mechanisms in our region. As we speak, East Asia has developed an all-round, multi-level and all-dimensional pattern of cooperation with ASEAN at its core, ASEAN-China cooperation as the basis, ASEAN-China-Japan-ROK cooperation as the main channel, China-Japan-ROK cooperation as the support, and East Asia Summit as the platform, demonstrating a sound momentum of vigorous development.

East Asian cooperation aims to achieve regional economic integration. Economic and trade integration in East Asia are making steady progress, with trade volume among China, Japan and the ROK approaching US$700 billion in 2014, and that between China and ASEAN reaching US$420.1 billion in the first 11 months of 2015. China and the ROK formally signed a free trade agreement in June last year, and negotiation on a China-Japan-ROK FTA is steadily moving forward. China and ASEAN successfully concluded negotiations on an upgraded FTA at the end of last year. Negotiations on the Regional Comprehensive Economic Partnership (RCEP) have accelerated, with all parties striving to conclude negotiations in 2016 and build an FTA with the largest population, the most diverse membership and the most dynamic development in the world.

East Asian cooperation aims to build a community with a shared future in the region. East Asian countries have set out the important goal of completing the East Asia Economic Community by 2020. ASEAN, which is at the center of East Asian cooperation, announced the establishment of the ASEAN Community at the end of last year. China always takes ASEAN as a priority in its neighborhood diplomacy, and firmly supports ASEAN integration and ASEAN centrality in regional cooperation. China and ASEAN have actively implemented the "2+7 Cooperation Framework" and promoted the formulation of the third plan of action, which provides impetus for building the China-ASEAN community with a shared future. Last year, the Lancang-Mekong Cooperation mechanism was launched upon China's initiative, and its first leaders' meeting will be held in conjunction with the annual meeting of the Boao Forum for Asia in 2016, striving to build a community with a shared future for Lancang-Mekong countries featuring mutual benefit and win-win cooperation.

East Asian cooperation aims to promote common development of all countries. East Asia is on the whole in the stage of rapid modernization, industrialization and urbanization, with huge demand for infrastructure connectivity and production capacity cooperation. China has taken an active part in the master plan on ASEAN connectivity and, on that basis, worked with the parties concerned to explore the formulation of a master plan on East Asia connectivity and advance the construction of the Trans-Asian Railway. Focusing on construction machinery, electric power, building materials, telecommunications and industrial parks, China is carrying out international production capacity cooperation in line with the need of ASEAN countries. At the China-Japan-ROK Trilateral Summit which resumed after three years, the three countries agreed to strengthen production capacity cooperation on manufacturing and services in the global market. East Asian countries have also deepened cooperation on agriculture-related poverty alleviation, personnel training, science, technology and education in order to deliver more benefits to the people and enhance capacity for sustainable development.

East Asian cooperation aims to enhance regional financial stability. To overcome financial risks in the region, it is imperative to develop the Asian financial system. All parties are determined to strengthen macro-policy coordination and jointly uphold economic and financial stability in the region. China has facilitated the establishment of the Asian Financial Cooperation Association jointly sponsored by regional financial institutions to strengthen exchanges among regional financial institutions and integration of financial resources. China has worked actively with other parties to enhance the effectiveness and operability of the Chiang Mai Initiative Multilateralization (CMIM), to give full play to the role of the ASEAN+3 Macroeconomic Research Office (AMRO), to promote the opening and development of the Asian bond market, to strengthen exchange and cooperation in the field of credit, and to build a robust network for financial security in the region.

This year is crucial for the full implementation of the Belt and Road Initiative. It is the first year of China's 13th Five-Year Plan, the 25th anniversary of dialogue relations between China and ASEAN, and the year of opportunities for East Asian cooperation. At the same time, our region also faces the challenges of economic downturn and structural transformation, and the most complex economic situation since the global financial crisis in 2008. We must remain confident, seize opportunities, overcome challenges, and firmly grasp the historical trend of the rise of Asia.

We must be confident in Asia's development. Asia remains the most dynamic region for world economic development. China, India and other major Asian economies are expected to maintain high growth rates for a long time, and the establishment of the ASEAN Community will further stimulate growth in Asia. According to the analysis on the global economy conducted by the International Monetary Fund, the Asian Development Bank, the *Economist* magazine as well as Goldman Sachs and other investment firms, Asia, which contributed to more than half of global growth, with about 30 percent coming from China, remains the best-performing region and the main driving force for global growth.

We must be confident in the Chinese economy. This year, the Chinese economy has withstood downward pressure and maintained its performance in a reasonable range despite the complex domestic and international situation. China's GDP grew by 6.9 percent in the first three quarters, which is still among the highest of major global economies. In the course of transformation and upgrade, the Chinese economy has stayed on track for long-term growth. The Chinese government has ample macro policy tools in its reserve and rich experience in addressing risks. We have the ability and confidence to maintain medium-high growth and move towards medium-high level of development.

China and its neighboring countries are linked by mountains and rivers, with closely-intertwined interests and a shared future, and always stand with each other in the face of challenges. To overcome barriers and achieve prosperity, we need to uphold the spirit of solidarity and foster the habit of cooperation. We must seize the opportunity of the Belt and Road development and East Asian cooperation, and increase the breadth and depth and raise the level of cooperation in our region. In this connection, I propose we strengthen cooperation in the following areas:

First, synergizing development strategies to achieve policy connectivity. Based on the principles of mutual respect, consensus and accommodating each other's comfort level, we need to enhance synergy between the Belt and Road, the national development plans of various countries and ASEAN's Post-2015 Vision, with the view to expanding consensus, identifying each other's needs and achieving win-win outcome.

Second, advancing infrastructure and production capacity cooperation to achieve infrastructure connectivity. We need to strengthen connectivity in areas such as road transportation, basic industries, economic cooperation zones and institution building, and vigorously develop a more accessible, safe and integrated transportation network, laying a solid foundation for common development.

Third, promoting free trade to achieve trade connectivity. We need to speed up negotiations on the China-Japan-ROK FTA and the RCEP, and seize the opportunity of the launch of the upgraded China-ASEAN FTA to improve regional trade structure and promote balanced growth of trade.

Fourth, strengthening financial cooperation to achieve financial connectivity. We need to build the Asian monetary stability system, investment and financing system and credit system, support the AIIB and the Silk Road Fund on the first batch of projects, and provide funding support to the Belt and Road development.

Fifth, deepening cultural exchanges to achieve people-to-people connectivity. We need to leverage the advantages of geographic proximity and cultural connections among regional countries, promote exchanges in education, culture, tourism, youth and other fields, carry out dialogue among civilizations, and organize the China-ASEAN Year of Educational Exchange activities to deepen friendship among our peoples.

As a special administrative region of China, Hong Kong has unique economic, geographic and human resources advantages. It is an international financial, trade and shipping center and the largest offshore RMB center. Given its superior geographical location, Hong Kong also serves as an important hub connecting the Mainland and overseas. It has an open economy, a good knowledge of international rules, and high-standard professional services such as finance, trade and logistics. It has a sound legal system and is building a regional legal service and dispute resolution center. Hong Kong can surely make use of its advantages as a super contact and play an important role in the Belt and Road development.

Hong Kong and ASEAN have close economic and trade relations, with ASEAN being Hong Kong's second largest trading partner in goods and the fifth largest FDI destination. As an independent entity, the Hong Kong Monetary Authority has actively participated in the CMIM process within the framework of ASEAN and CJK cooperation. Hong Kong and ASEAN are negotiating an FTA, which will be conducive not only to closer cooperation between Hong Kong and ASEAN, but also to the process of regional cooperation.

As an important economic entity, Hong Kong's participation in the Belt and Road development and regional economic cooperation will bring opportunities for its own development, expand its space for external cooperation, and contribute to its economic transformation. It will help Hong Kong maintain long-term prosperity and stability, and make this oriental pearl shine even brighter in the new century.

The Belt and Road is our common cause, and East Asian cooperation is a process with a promising future. We welcome the Hong Kong SAR government and friends from all sectors to take part in this cooperation process, and make a joint effort toward a brighter future.

Deepening Regional Cooperation for a Community of Shared Future

Speech at the "Asian Regional Cooperation Roundtable" of the Boao Forum for Asia Annual Conference 2016

March 25, 2016

Regional cooperation in Asia has come a long way. In modern times, as Asian countries fought for national independence and freedom from poverty and backwardness, people with foresight had already started to explore ways for regional cooperation in Asia in light of their countries' role and standing and that of Asia in the evolving international landscape. The Bandung Conference in 1955 gave a strong boost to the effort of Asian countries to gain strength through unity. The post-Cold War years saw a new boom in regional cooperation in Asia. As history demonstrates, harmony benefits all Asian countries while confrontation hurts all. The rise of Asia has not come easily and unity remains the best way to build a stronger Asia.

Today, Asia is a rising region in the global geopolitical landscape, and is also leading world economic growth. Regional cooperation now enjoys a strong momentum and is blossoming into a comprehensive architecture that covers all sub-regions and consists of mutually complementary frameworks. This has become a hallmark achievement of Asia. The ASEAN community set up at the end of last year is a milestone in regional integration. East Asian cooperation has evolved into a multi-tier and all-dimensional framework with ASEAN at the center, 10+1 cooperation as the basis, ASEAN plus three cooperation as the main channel, China-Japan-ROK cooperation as a major component and East Asia Summit as an important platform. Besides these mechanisms, the South Asian Association of Regional Cooperation (SAARC) plays a significant role in promoting peace, stability and development of South Asia. The Shanghai Cooperation Organization (SCO) is a constructive and important force in the Eurasian geopolitical structure. The Asia Cooperation Dialogue (ACD), the only official dialogue and cooperation mechanism open to all Asia, has been active in deepening cooperation and friendship among Asian countries. The Conference on Interaction and Confidence Building Measures in Asia (CICA) is another important venue for promoting confidence, dialogue and cooperation. The Asia-Europe Meeting (ASEM), ASEAN Regional Forum (ARF) and East Asia Summit (EAS) have drawn in members outside of Asia, showing their openness and inclusiveness. Deepening and expanding cooperation within these cooperation frameworks has provided a strong impetus for regional integration and the rise of Asia.

Cooperation in Asia has deep roots in history, and reflects the trend of the times and the need of the region. In summary, Asian regional cooperation has come this far for the following reasons:

First, it builds on the trend of the region and the world. In this increasingly multi-polar, globalized and digitized world, countries are more interdependent in their interests than ever before. No country can develop on its own. The importance of common development and seeking strength through unity has been widely recognized. Regional integration is becoming the overall trend. In this area, Europe is the frontrunner, and North America, Latin America and Africa are also making similar efforts. It is a natural choice for Asian countries to promote their version of regional integration.

Second, it serves the common interests of regional countries. Peace and development are the mainstreams in Asia. They serve the common interests of countries and peoples. Asian countries loathe seeing a repeat of war, turmoil and mutual hostility, and are eager to focus on development in a peaceful and stable environment. Asian countries are deriving strength from complementarities and are engaged in a concerted effort to deepen regional cooperation.

Third, it is based on a unique and effective Asian approach. Due to its diversity, Asia cannot copy the model of others in seeking regional integration. In the course of its development, ASEAN has evolved a set of principles called "the ASEAN way," which include mutual respect, equality, consensus-building and accommodating all parties' comfort level. Today, the "ASEAN way" has been widely seen as a successful model of cooperation and an important way to handle relations among Asian countries. This has contributed to the fostering of a new type of international relations.

Fourth, it embodies the principle of openness and inclusiveness.

Asian cooperation is neither a small club nor an exclusive process. Over the years, Asian cooperation frameworks have looked to, engaged with and learned from those in other regions. They also provide important platforms for parties outside Asia to participate in and play a constructive role in Asian cooperation. This is conducive to positive interactions and common development.

While encouraged by fast progress, we should stay cool-headed at the difficulties and challenges. Asia still faces serious traditional and non-traditional security threats. Given the prolonged weakness in world economic growth, rising financial risks and lower commodity prices, Asian countries are under mounting pressure in advancing reform, fending off risks and achieving growth. Pan-Asia cooperation remains just an academic concept. And security cooperation is still a weak link. After fast expansion over the past two to three decades, the low-hanging fruits have been all but harvested and we have entered the deep water zone in regional cooperation. It requires greater political resolution, courage and wisdom from all parties to tackle the difficulties and advance cooperation.

China is an active participant in Asian regional cooperation and a steady contributor to building a community of shared future. Since becoming a dialogue partner with ASEAN 25 years ago, China has seen its participation in regional cooperation grow both in breadth and depth. China and ASEAN countries are building a closer community of shared future and had finished the negotiation on upgrading the free trade area last year. China supports the building of an East Asian Economic Community among China, Japan, the ROK and ASEAN by 2020 and supports speeding up negotiations on Regional Comprehensive Economic Partnership (RCEP). Just two days ago, the first Leaders' Meeting of Lancang-Mekong Cooperation (LMC) was successfully held in Sanya. As a new sub-regional cooperation framework, it will promote the development and prosperity of the sub-region and contribute to the ASEAN community building and its integration. China strongly supports the SCO and actively participates in ACD, ASEM, APEC and other mechanisms. China is now advancing the Belt and Road Initiative. It has initiated the Asian Infrastructure Investment Bank (AIIB) and most recently proposed the setting up of an Asia Financial Cooperation Association. All these have contributed to regional integration and lent fresh impetus to regional cooperation.

As a Chinese saying goes, "Even with great success one should always try to do better." China is ready to work with other Asian countries to take regional cooperation to new areas and higher levels. In this context, I would like to share a few suggestions on the way forward.

First, to consolidate the foundation. Regional cooperation is an important means to improve regional governance and fuel the rise of Asia. We should stay committed to win-win cooperation and common development, make regional cooperation a priority in our national development and foreign policy agenda, and increase constructive interactions among regional countries. Meanwhile, we should work to enhance mutual trust, bridge gaps, narrow differences and build consensus so as to consolidate the foundation and create an enabling atmosphere for deeper regional cooperation and integration.

Second, to give priority to development. Development is still the key to solving the major issues and problems in Asia. Asian countries should make full use of regional cooperation platforms to share experience on reform and innovation, urbanization, industrialization, poverty reduction and livelihood improvement, strengthen connectivity and achieve inclusive and integrated development. We should continue to pursue open regionalism, ensure mutually reinforcing and coordinated development among various trade arrangements and cooperation frameworks and build an open, integrated, balanced and win-win regional economic architecture.

Third, to improve institutional building. Sound institutions could help to guarantee the deepening of cooperation. We should explore ways to enrich and improve regional and sub-regional cooperation mechanisms. The goal is to make sure they have distinctive focus and features while remaining open and inclusive. Together, they could form an efficient cooperation framework with full functions, clearly defined and complementary priorities. This would better meet the expectations of regional countries for security, development and friendship through cooperation.

Fourth, to maintain Asian features. Both the content and pattern of regional cooperation should be based on Asia's reality and serve its peaceful development. We need to continue with the Asian way of mutual respect, consensus-building and accommodating all parties' comfort level. We should keep up with the times and be innovative, explore new ideas and new ways in light of the trend of Asia and people's needs and further elevate cooperation, tackle difficulties and make cooperation more effective.

Advance Island Economic Cooperation-A New Platform for Regional Cooperation

Remarks at the "21st-Century Maritime Silk Road: Islands Economic Cooperation Sub-Forum" of Boao Forum for Asia Annual Conference
March 25, 2016

President Xi Jinping put forward the initiative of jointly building the Silk Road Economic Belt and 21st Century Maritime Silk Road in 2013. Since then, the initiative has seen endorsement and active participation of over 60 partner countries. The Belt and Road Initiative (BRI) follows the principle of consultation and cooperation for shared benefits, and advocates the spirit of openness, inclusiveness, cooperation and mutual benefits. It carries forward the tradition of friendship among different civilizations over the past hundreds or thousands of years, and points to a future of common development and prosperity for the partner countries, providing an effective platform for cooperation among countries. From the South China Sea, the 21st Century Maritime Silk Road extends westward to the Mediterranean and southward to the South Pacific. Large numbers of islands lie in important locations in these areas. Situated on key routes for international business, commerce and personnel exchanges, these islands will have an important role to play in cooperation on the 21st Century Maritime Silk Road.

Under the influence of economic globalization and by taking the unique advantage of their locations, industries, and cultures, some island countries and regions have achieved leap-frog development in a short period of time and become famous "pearls in the sea". Among them are financial and trade centers like Hong Kong, Singapore as well as famous tourist destinations like Bali, the Maldives, Fiji, and Malta. Their various development models provide inspiration and experience for other island economies. Their useful experience includes making development strategies by fully leveraging external opportunities, enhancing infrastructure, promoting the free flow of economic factors, and vigorously developing tourism and cultural industry.

In building the 21st Century Maritime Silk Road, partner countries need to fully tap the potential of their resources and promote policy coordination, facilities connectivity, unimpeded trade, financial integration, and bonding among the peoples. The endeavor will generate a host of opportunities for the prosperity of island economies.

First, opportunities will come from the synergized development strategies among countries under the BRI cooperation. Island economies are endowed with enormous marine resources. That said, however, given the size of their territory and population, island economies have to depend on external markets to turn such resources into drivers of economic growth. The BRI advocates cooperation between partner countries to synergize development strategies and make plans for regional cooperation. China has signed MOUs or intergovernmental documents on the BRI with over 20 partner countries and sought to align the 21st Century Maritime Silk Road with Indonesia's Global Maritime Fulcrum as well as Malaysia's Economic Transformation Program. We will strengthen communication and form synergy with more island economies to share in the opportunities offered by China's development and regional cooperation and help them fully tap their potential for leap-frog development.

Second, opportunities will come from infrastructure connectivity that BRI aims to achieve. Only by building better infrastructure and maritime arteries can island economies improve their geo-economic value and turn oceans' gifts, that is, marine resources, into endless financial revenues. Under the 21st Century Maritime Silk Road, China has worked with Indonesia, Malaysia, Myanmar, Sri Lanka, Pakistan, among others, to launch port infrastructure construction projects and develop supporting industries with major progress. We look forward to having more island economies joining in and building maritime arteries for greater connectivity and inclusive and interconnected development.

Third, opportunities will come from trade and investment facilitation which is a key component of the BRI. The free flow of goods and capital is crucial for island economies to enhance economic vitality and promote industrial development. Under the 21st Century Maritime Silk Road, China has signed free trade agreements with partner countries and regional organizations, including Singapore, New Zealand and ASEAN. China has also deepened industrial capacity cooperation with Indonesia, Malaysia, and Brunei to help diversify industries and improve their industrial structure. In the spirit of reciprocity and mutual benefit, China will make closer arrangements of trade and investments with partner island economies to jointly create a full-fledged and diversified marine industrial chain and value chain, unleashing the potential for expanded practical cooperation.

Fourth, opportunities will come from the new investment and financing channels that the BRI aims to provide. Some island economies rely on foreign financial resources and face large funding gaps in their infrastructure development. Their export revenues and overseas financing costs have been affected by volatility in the international financial market. The BRI is committed to creating a regional investment and financing system. Over one-sixths of the 57 member states of the Asian Infrastructure Investment Bank are island economies. China will continue to push forward the process of RMB internationalization, gradually expand the scale of currency swap and settlement with partner countries, and support partner countries in issuing RMB-denominated bonds, creating new platforms for island economies to expand the financing channels and defuse financial risks.

Fifth, opportunities will come from the tourism boom that the BRI will bring. In many island economies, turning tourism into a pillar industry and enhancing its cultural dimension is the top development strategy. The BRI advocates stronger cultural exchanges and tourism cooperation. In 2015, the total number of China's outbound tourists exceeded 140 million, and many island economies became popular tourist destinations. Take Palau, the island country in the South Pacific, for instance, half of its foreign tourists were from China's Mainland. China will promote cooperation on cruise tourism under the 21st Century Maritime Silk Road. Hence partner island economies will be able to bring into full play their strengths in tourism and share in the development opportunities.

Actively Practice the Asian Security Concept and Jointly Create a New Future of Asia-Pacific Security

Remarks at the opening ceremony of the International Seminar on Security Framework and Major-Power Relations in the Asia-Pacific Region
July 12, 2016

In recent years, the security architecture and major-country interactions in the Asia-Pacific have attracted wide attention. With this as the theme of discussion, participants will exchange views, pool wisdom and offer suggestions on the security concept, security agenda, cooperation mechanism as well as major-country relations in the Asia-Pacific. The meeting is therefore timely and relevant.

An ancient Chinese saying goes: "A gentleman is constantly mindful of danger, fatality and chaos even when he is in a state of security, survival and order. And this ensures the security of himself and the state." Indeed, since ancient times, mankind have been looking for ways to walk out of the security dilemma and reach the state of common security of ensuring security of oneself without damaging the security of others.

From the vision of a world of harmony and justice to the idea of benevolence that promotes righteousness and rejects hegemony, from Plato's Republic to Immanuel Kant's perpetual peace, and from Woodrow Wilson's "Fourteen Points" to the Atlantic Charter, mankind's quest for peace and justice has never ceased. From the Westphalian system, the origin of modern international relations, to the Vienna system in the 19th-century Europe, from the post-World War I Versailles-Washington system, the post-World War II Yalta system, to the Helsinki process which shaped common security in Europe, the efforts to build a sound security architecture have never stopped.

Now, let's turn our eyes back to the Asia-Pacific and take a look at the building of security architecture in our own region. Currently, there are five types of security mechanisms in the Asia-Pacific: First, the US-led alliance system and relevant bilateral and multilateral arrangements; Second, the ASEAN-centered security dialogue and cooperation mechanisms such as the ASEAN Regional Forum (ARF) and ASEAN Defense Ministers' Meeting-Plus (ADMM+); Third, special mechanisms on hotspot issues such as the Six-Party Talks and the Quartet on Afghanistan; Fourth, cross-region security cooperation mechanisms such as the Shanghai Cooperation Organization (SCO) and Conference on Interaction and Confidence-Building Measures in Asia (CICA); Fifth, Track 1.5 or Track 2 security dialogue platforms involving participants from the region and beyond such as the Shangri-La Dialogue, Xiangshan Forum and the Asia-Pacific Roundtable.

Since the end of the Cold War, the Asia-Pacific has avoided major conflicts and wars and maintained overall peace in a complex and volatile environment. This is not God's blessing or historical coincidence, but the result of joint efforts of all Asia-Pacific countries and the work of various security mechanisms under different conditions at different times. Security architecture in the Asia-Pacific on the whole needs improvement.

Today, the Asia-Pacific is witnessing a faster adjustment of geopolitical structure while enjoying overall stability. On the one hand, the tug-of-war between major countries has escalated. The Korean nuclear issue, Afghanistan and other hotspot sensitive issues are heating up. Territorial and maritime disputes are thrown into sharper relief. Terrorism, climate change, natural disaster, humanitarian crisis and other non-traditional security issues are on the rise. These have posed more complex and grave security challenges to the Asia-Pacific. On the other hand, the balance of power in the region has seen major changes. China and other developing countries are growing stronger and playing an even more constructive role in regional and international affairs. We need reasonable development space, an appropriate say in the world, and a new regional security structure that is reflective of the reality of the region, consistent with the interests and needs of various parties and acceptable to all.

In this context, countries have put forward various initiatives and proposals, for example, the rules-based Asia-Pacific security system by the US, the Asia-Pacific regional security and cooperation structure by Russia, the ASEAN's call for further consolidation of ASEAN centrality, the dialogue-based Asia-Pacific security architecture by India, the

Northeast Asia Peace and Cooperation Initiative by the ROK, and the list goes on. What kind of regional security architecture to build and how to build have been popular subjects of discussion among the parties concerned.

As an important member of the Asia-Pacific family, China has all along been a builder of peace, contributor to development and defender of security in the Asia-Pacific. China is ready to contribute its wisdom and proposals to maintaining peace and stability in the region. Here, let me share with you some of my observations on the building of Asia-Pacific security architecture.

First, the architecture should be guided by the goal of common security and common development. Security and development, like two wings of a bird and two wheels of a bike, are mutually reinforcing and indispensable. It is important to ensure common security and pursue common development through the building of the Asia-Pacific security architecture.

In the 21st century, Asia-Pacific countries have formed a community of shared future and common responsibility. Confronted with complicated risks and challenges, countries cannot uphold their own security by themselves. Pursuing a beggar-thy-neighbor policy, forming alliance or opting for confrontation only escalates tensions. This makes cooperation the only way out. We should be firmly committed to win-win cooperation, and reject the outdated mindset of zero-sum game or "the winner takes all." We should accommodate the interests of others while upholding our own interests, promote common development while seeking our own development, and follow a path of common security underpinned by dialogue, consultation and win-win cooperation. The building of the Asia-Pacific security architecture should not be dictated by a single country or a few countries. And the architecture should meet the common security interests of all countries instead of any one country or a small group of countries. Take the move by the ROK and the US to deploy the THAAD system in the ROK as an example. They have only considered their own security with no regard to the security of others. But ultimately, their own security may not be ensured in this way.

Second, the architecture should be based on widely acceptable norms governing international relations. Ideas guide actions. In this case, the concept of Asia-Pacific security should be constantly updated to guide the development of the security architecture.

The architecture should be fostered with three principles: Number one, firmly uphold the UN-centered post-war international system and basic norms governing international relations, particularly the purposes and principles of the UN Charter. Number two, observe universally recognized international law and international rules, such as the Five Principles of Peaceful Coexistence, the United Nations Convention on the Law of the Sea (UNCLOS), the Treaty of Amity and Cooperation in Southeast Asia (TAC), and the Declaration on the Conduct of Parties in the South China Sea (DOC). Number three, endorse and embrace the ASEAN Way of equality, consensus and accommodating each other's comfort level, based on the common understanding reached among regional countries in their exchanges and interactions. Only by following such principles and spirit, can we make steady progress in pursuing a path of Asia-Pacific security explored, built, shared and protected by all, and solidify the foundation for the Asia-Pacific security architecture.

Third, the architecture should be strengthened by building partnerships. Those who share the same ideal and follow the same path can be partners. Those who seek common ground while shelving differences can also be partners. Asia-Pacific countries should work together to take a new path which chooses dialogue over confrontation, partnership over alliance, and amity over grudges, and jointly foster an Asia-Pacific partnership of mutual trust, inclusiveness, mutual benefit, reciprocity, mutual understanding and mutual accommodation.

Whether effective security architecture can emerge in the Asia-Pacific will depend on how China and the US interact with each other in this part of the world. Both important members of the Asia-Pacific, China and the US shoulder great responsibilities for peace and prosperity in the region. The two countries are building a new model of major-country relations featuring no conflict or confrontation, mutual respect and win-win cooperation. This is in line with the common expectation and aspirations of the countries in the region as well as the fundamental interests of themselves. The two sides need to develop a correct understanding of each other's strategic intention. China does not want to be the predominant power in the Asia-Pacific, or build spheres of influence and military alliance. It has no intention to replace the US or exclude the US from the region. And it does not seek to be part of the so-called G2. We hope the US will play a constructive role in the Asia-Pacific, recognize and accept the fact of a growing China, cast aside the

Cold War mentality, respect China's sovereignty and territorial integrity, and coexist with China peacefully. The US "rebalance" to the Asia-Pacific should not be advanced on ideological grounds, turning this region into a showground of naval and air power, or forcing regional countries to take sides. The relevant bilateral military alliances are a product of a bygone era. In this new age, they should be more transparent and not target a third party or aim at holding back and blocking China.

Parties should continue to support the development of the mechanisms in which ASEAN plays a leading role, respect its centrality in regional cooperation, and demonstrate such support and respect with real actions. China welcomes Russia's interest and input in the Asia-Pacific, and wants to see Russia and regional countries working together as a stabilizer for peace and security in the region. China also attaches great importance to further strengthening China-Japan-ROK trilateral cooperation. Other parties also need to take actions to play a constructive role and jointly put in place Asia-Pacific security architecture.

Fourth, the architecture should take the form of an integrated, multi-layered, and diversified network. The architecture we are building is not a brand new project that dismisses all previous arrangements, but an upgrading and improvement of existing ones. It should be a multi-layered, broad-based, and integrated network that is open, inclusive, cooperative, and deliver win-win outcomes. It is based on existing security mechanisms and open to countries outside the region. It should be an anchor for regional peace and stability.

The current mechanisms in the region must be updated to play a bigger role. Parties should support the "parallel-track" approach of advancing denuclearization of the Korean Peninsula and replacing the armistice agreement with a peace treaty. Parties concerned should increase dialogue, resume the Six-Party Talks at an early date, and ensure long-term peace and stability in Northeast Asia. Various parties should continue to promote peace and reconstruction in Afghanistan and support an Afghan-led and Afghan-owned inclusive political reconciliation process. The Shanghai Cooperation Organization should play a bigger role; the Conference on Interaction and Confidence-Building Measures in Asia should strengthen its institutional building; and regional security dialogue platforms, such as the Xiangshan Forum and the World Peace Forum, should achieve greater development. We must take an active part in the dialogue and cooperation under various arrangements, improve the relevant institutions, put forward more initiatives and proposals that meet the needs of regional countries, and provide more public goods, in order to gradually put in place an Asia-Pacific security architecture and system that is consistent with the trend of the times and the needs of countries in the region and encourages broad participation.

The world today has better conditions than ever before to pursue peaceful development. As a responsible major country, China is committed to the path of peaceful development and regards promoting world peace and development its sacred responsibility. Meanwhile, it is only natural that China will firmly uphold its sovereignty, security and development interests. No one should expect China to live with the consequences of having these compromised.

For China, Asia-Pacific is its home and foundation for development and prosperity. China calls for an Asian security concept featuring common, comprehensive, cooperative, and sustainable security. It stands for consultation and dialogue, no threat of force; openness and inclusiveness, no mutual exclusion; and win-win cooperation, no zero-sum game mentality. Such a concept opens up broad prospects for regional security cooperation. China actively participates in bilateral and multilateral security dialogue and cooperation, regional security frameworks, and cooperation in non-traditional security fields, such as disaster prevention and reduction, search and rescue, counter-terrorism, and combating transnational crimes. China is stepping up policy coordination with other countries on the management of hotspot issues and regional security cooperation, promoting the development of a common security code of conduct, and providing new impetus for building security architecture in Asia-Pacific through innovations in security concepts and practices.

China is implementing the Belt and Road Initiative first in Asia, seeking synergy between its development strategies and those of neighboring countries, and advancing cooperation in infrastructure and production capacity. The Asian Infrastructure Investment Bank has been launched. We are strengthening regional cooperation institutions, including 10+1, 10+3, China-Japan-ROK, APEC and the SCO, and building a more open, inclusive, and mutually beneficial network of regional cooperation. We are taking forward Regional Comprehensive Economic Partnership (RCEP) negotiations, developing the Free Trade Area of the Asia-Pacific (FTAAP), and exploring ways to connect various free trade arrangements for better coordination and complementarities. We are deepening regional cooperation and economic integration to provide solid economic and social support for the Asia-Pacific security architecture.

For some time, the situation in the South China Sea has been a focus of attention. In a few days, the arbitral tribunal on the South China Sea initiated by the Philippines will issue the so-called final "award." So let me say a few words about the South China Sea issue.

The Nansha Islands are an integral part of Chinese territory. This is supported by a mountain of historical and legal evidence. It is only right and natural for a country to stand up for its territorial sovereignty. This arbitration case is in fact a political provocation of China's sovereignty and gravely violates the principles of international law. The arbitral tribunal has turned a blind eye to the real nature of China-Philippines disputes, which is territorial and maritime delimitation, ignored China's declaration under Article 298 of UNCLOS, and arbitrarily heard and exercised jurisdiction on a case that clearly involves matters beyond UNCLOS. This is a willful attempt to expand and act beyond its authority. The arbitration case has been built on an illegal basis from the very beginning. Naturally its so-called "award" is also illegal. China does not accept or participate in the case, and will not recognize its verdict. This is to defend our rights, and also uphold international rule of law.

The only way to settle the South China Sea disputes is for countries directly concerned to carry out negotiation. Historical experience shows that negotiated solutions to disputes not only reflect the independent will of the countries concerned and sovereign equality, but also stabilize and promote bilateral relations. China and the Philippines are neighbors that cannot be moved away. Good-neighborliness and friendship are historical traditions of our two countries, and the right way to go. We hope the new administration of the Philippines will adopt a new approach to the South China Sea, bring the disputes to the negotiating table, and take China-Philippines relations back to the track of development.

To safeguard peace and stability in the South China Sea is a common responsibility of China and other countries in the region. China stands ready to address the South China Sea issue with the "dual track" approach that has won extensive recognition from countries in the region, namely peacefully resolving relevant disputes through friendly consultation among countries directly concerned and upholding peace and stability in the South China Sea by China and ASEAN countries working together. China is committed to implementing the DOC and working for an early conclusion of the COC. China will, as always, make joint efforts with all parties to uphold peace and stability with rules and mechanisms, achieve win-win outcomes through development and cooperation, and ensure that all countries enjoy freedom of navigation and over-flight according to law, so as to make the South China Sea a sea of peace, friendship, and cooperation.

Deepen Production Capacity Cooperation in Jointly Building the Community of Shared Future

Speech at the second session of the 21st Century Maritime Silk Road and Promoting International Production Capacity and Equipment Manufacturing Cooperation Forum

Nanning, September 11, 2016

This year is the first year of the ASEAN Community as well as the 25th anniversary of China-ASEAN dialogue relations, so it bears historic significance. Over the past 25 years, holding high the banner of development and cooperation and pursuing mutual respect, understanding, trust and support, China and the ASEAN have become sincere partners with shared future, mixed interest and profound friendship. Recently, China-ASEAN cooperation has witnessed fruitful highlights. We worked together in the establishment of the Asian Infrastructure Investment Bank, completed negotiations on upgrading the China-ASEAN Free Trade Area, actively implemented the Belt and Road Initiative and set up the Lancang-Mekong River cooperation mechanism. In 2015, trade volume between the two sides reached US$472.2 billion and China was the biggest trade partner of the ASEAN for the seventh consecutive year. The accumulative volume of mutual investment between the two sides exceeded US$160 billion and the number of annual personnel exchanges exceeded 23 million. China-ASEAN relations have become the cornerstone of peace and stability, the driving force of development and prosperity and an example of cooperation in the region. Four days ago, Premier Li Keqiang and leaders of the ten ASEAN countries successfully held the Commemorative Summit Marking the 25th Anniversary of China-ASEAN Dialogue Relations. During the summit, leaders from China and ASEAN countries gathered together to deepen friendship and discuss cooperation, setting goals and charting the course for the future development of bilateral relations. China-ASEAN relations are going to embark upon a better future at the new starting point.

As a key area of China-ASEAN cooperation, production capacity cooperation is also an inherent part of the joint building of the Maritime Silk Road between China and neighboring countries. Joint Statement on Production Capacity Cooperation among Lancang-Mekong countries was issued at the first Lancang-Mekong Cooperation Leaders' Meeting in Sanya this March and Joint Statement between China and ASEAN on Production Capacity Cooperation was issued at the Commemorative Summit Marking the 25th Anniversary of China-ASEAN Dialogue Relations, which serve as an important guide for production capacity cooperation between China and ASEAN countries. At present, bilateral production capacity cooperation is in the ascendant, and preliminary results have been achieved in a batch of infrastructure development and industrial park projects. This is mainly attributed to the following reasons:

First, it complies with the economic transformation and upgrade of the two sides. Currently, Asian countries are coping with the mounting downward economic pressure through structural reform. China is stepping up its efforts in promoting economic transformation and upgrade, strengthening supply-side structural reform, and accelerating the development of the new economy. ASEAN countries are also deepening economic restructuring and forging new investment and growth points. Against the new situation, strengthening international production capacity cooperation is an important move to maintain the medium-high growth rate and strive towards a medium-high level of the Chinese economy, which is also an important part of the new round of high-quality opening-up. At the same time, it will also improve the economic quality and efficiency of ASEAN countries and advance the robust and sustainable development of the Asian and world economy.

Second, it is in line with the development needs and interest appeals of Asian countries. Asian countries, many of which are in the initial stage of industrialization and urbanization, give top priority to development. They have demands and willingness to strengthen infrastructure development, advance connectivity, and carry out international production capacity cooperation. After years of development, China has entered the mature stage of industrialization with a large amount of competitive industries. We are willing to deepen production capacity cooperation with Asian countries including ASEAN countries, share experience in industrial civilization and peaceful development, help relevant countries improve infrastructure development, elevate the level of industrialization, speed up the development of the real economy, and enhance their position in the global industrial chain and value chain.

Third, it follows the unique and effective Asian way. Asian countries have established the Asian way of mutual respect, equal treatment, consensus through consultation and accommodation of the comfort level of all sides through years of interaction and communication. It reflects the core of the Asian way and highly fits with the Silk Road spirit when China carries out production capacity cooperation with Asian countries on the basis of voluntariness, equality, and mutual benefit, thus winning positive support and response from various countries. China is ready to further dovetail development strategies, reinforce production capacity cooperation in a pragmatic and effective manner, and achieve mutual benefit and win-win results to the greatest extent with other countries. We will strictly abide by market principles, pursue international norms, and uphold openness and inclusiveness in cooperation as well.

International production capacity cooperation is a systematic project, and the Maritime Silk Road is a long-term project. China is willing to work with ASEAN countries and countries along the Maritime Silk Road to deepen production capacity cooperation so as to inject impetus into respective development and mutually beneficial cooperation. In this regard, I would like to share a few ideas with you.

Firstly, we should deepen China-ASEAN relations. In the next 25 years, China is willing to, together with the ASEAN, firmly grasp the general direction of the development of bilateral relations, further strengthen strategic communication, enhance the docking of development strategies and paths, propel the implementation of the "2+7 cooperation framework", and usher China-ASEAN relations from the "growth stage" of rapid development into the "mature stage" of quality improvement and upgrade so as to forge an even closer China-ASEAN community of shared future and let the cooperation results better benefit regional countries and their people.

Secondly, we should strive to build highlight projects. We should continue to assume a cooperation attitude of pragmatism and efficiency, and make steady headway in the development of connectivity infrastructure projects, such as China-Laos Railway, China-Thailand Railway, and the Jakarta-Bandung High-speed Railway, so as to form the Trans-Asia Railway Network at an early date and set an example for regional connectivity. We must further plan the construction of China-Laos industrial park, China-Malaysia industrial park, China-Myanmar industrial park, China-Indonesia industrial park and other industrial parks, optimize the layout of the industrial chain, accelerate regional industrial upgrade, and serve regional development.

Thirdly, we should make the most of the mechanisms and platforms. We should give full play to the leading role of China-ASEAN Summit and take full advantage of such platforms as the China-ASEAN Expo to bolster the cluster development of production capacity cooperation. Lancang-Mekong Cooperation is a new highlight and growth point of China-ASEAN cooperation, and production capacity cooperation is a preferential field in Lancang-Mekong Cooperation. We must make full use of the Lancang-Mekong Cooperation mechanism, earnestly carry out early-harvest projects and improve sub-regional cooperation, so as to promote the integrated development of regional industrial chains.

Fourthly, we should accelerate the building of Maritime Silk Road. We should continue to deepen cooperation with ASEAN countries in marine economy, maritime connectivity, maritime security, maritime people-to-people and cultural exchanges, environment protection, scientific research and other areas, strengthen practical cooperation with Pakistan, Sri Lanka and other countries along the Maritime Silk Road, speed up the development of projects including the China-Pakistan Economic Corridor and the Bangladesh-China-India-Myanmar Economic Corridor, expand production capacity cooperation to wider fields and higher levels, and aid early harvest results from the Belt and Road development.

Work Together to Improve Regional Security Architecture and Address Common Challenges

Remarks at the First Plenary Session of the Seventh Xiangshan Forum in Beijing on October 11, 2016

Since the end of the Cold War 26 years ago, the trend of multi-polarity, globalization and regional integration have grown ever stronger. The Asia-Pacific has been peaceful and stable in general and is becoming the most dynamic region with the biggest potential. At the same time, our region faces increasingly complicated security issues, often diversified, emergent, transnational, and interlinked. Traditional hotspots flare up from time to time, and non-traditional security challenges are posing severe threats to the security of regional countries and regional stability.

In this context, the building of regional security cooperation architecture in the Asia-Pacific is lagging behind. Currently, there are five types of security mechanisms in this region. First, the alliance system led by the United States and relevant bilateral and multilateral arrangements; Second, the ASEAN-centered security dialogue and cooperation frameworks such as the ASEAN Regional Forum (ARF) and the ASEAN Defense Ministers' Meeting Plus (ADMM+); Third, special mechanisms on hotspot issues such as the Six-Party Talks on Korean Peninsula nuclear issue and the quadrilateral coordination mechanism on the Afghanistan issue made up of Afghanistan, China, Pakistan and the US; Fourth, regional security cooperation mechanisms including the Shanghai Cooperation Organization (SCO) and the Conference on Interaction and Confidence-Building Measures in Asia (CICA); Fifth, Track 1.5 or Track 2 security dialogues such as the Shangri-La Dialogue, Xiangshan Forum and the Asia-Pacific Roundtable. These security mechanisms reflect underlying disconnections in our region: problems left by the Cold War, lack of coordination among sub-regions, and differences on security concepts.

Economic cooperation and political and security cooperation, as two wheels driving Asia-Pacific cooperation, should complement each other and move forward in parallel. In the economic sphere, a relatively mature and stable framework has been cultivated to effectively promote regional economic integration. In the security area, in contrast, the fostering of security cooperation architecture has lagged behind, making it more difficult to deal with growing security challenges in a timely and effective way. This calls for the building of Asia-Pacific security architecture consistent with regional conditions and the interests of all parties.

In recent years, relevant parties have made valuable explorations in this regard and proposed some new visions and initiatives. At the Fourth CICA Summit in 2014, Chinese President Xi Jinping provided a Chinese vision, that is, to update our security concept, establish new regional security and cooperation architecture, and jointly chart a course for security that is by all and for all. This vision reflects the collective wisdom and consensus of regional countries and creates new prospect for security cooperation in the Asia-Pacific.

As an important member in the Asia-Pacific family, China has been a contributor to regional peace and defender of regional security. China is willing to work with all other parties to explore and build regional security architecture for the building of the Asia-Pacific community of common future. The new regional security architecture should have the following elements:

First, it should be guided by a new security concept. Actions are based on concepts. Security architecture that fits regional reality must be built on an updated security concept. Old security concepts, such as the mentality of Cold War and zero-sum game, are increasingly out of touch with the trend of globalization. Win-win cooperation has become the new spirit of the times. Tackling global challenges through common efforts is now the only viable choice for all countries. Asia-Pacific countries should enhance dialogues and cooperation, and explore new concepts and new approaches to safeguard regional security.

China has initiated the security concept featuring common, comprehensive, cooperative and sustainable security, which advocates consultation and dialogue, openness, inclusiveness, and win-win cooperation. This security concept is in line with the trend of our times. It is built on existing concept of regional security cooperation, and has injected new vitality to the development of Asia-Pacific security architecture. China is ready to promote and implement this new security concept together with other regional countries, and push forward the building of Asia-Pacific security architecture featuring wide consultation, joint contribution, shared benefits, and win-win outcome.

We believe that countries in the region should handle differences properly and peacefully through dialogues and friendly consultations and negotiations. This is in line with our regional norms of mutual respect, seeking common ground while shelving differences and peaceful coexistence. With efforts spanning 60 years, China has properly settled boundary issues with 12 out of its 14 land neighbors. The negotiations on boundary issue between China and India, and between China and Bhutan are moving forward steadily. With the joint efforts by China and ASEAN countries, the situation in the South China Sea has cooled down, and the regional rule and framework prevailed once again, which emphasizes the management, control and resolution of differences by consultations and negotiations.

Faced by heightened tension on the Korean Peninsula, China strongly urges the DPRK and other parties concerned to exercise restraint, truly comply with Security Council's resolutions and work hard to resume dialogue. Applying military pressure and undermining the strategic balance in the region will only deepen the security concerns of relevant parties and harm the prospect for resolving the Peninsula issue.

China actively supports and facilitates cooperation in non-traditional security areas. Every year we sponsor more than one third of the cooperation projects under the ARF, the East Asia Summit (EAS) and other regional frameworks, which has effectively promoted the non-traditional security cooperation in the region.

Second, the regional security architecture should be based on the rule of law and international norms. Without rules and norms, even a family would fall into disorder, not to say a country. Rule of law and norms are thus essential elements in the building of the Asia-Pacific security architecture. At the same time, the rules should be based on consensus and universally recognized international and regional norms. The will of a few countries cannot be equated to international or regional rules, nor taken as the sole basis for "a rules-based order." International law should not be interpreted out of context, and such interpretation should not be preached as so-called rules of law in international relations.

To be specific, the new Asia-Pacific security architecture should comply with the following rules: First, the UN-centered post-war international system and international order, and the fundamental principles of international law and basic norms of international relations enshrined in the UN Charter. Second, universally recognized rules of international law, including the Five Principles of Peaceful Coexistence, the 1982 UNCLOS and the Treaty of Amity and Cooperation in Southeast Asia. Third, codes of conduct jointly formulated by countries in the region, such as the Declaration on the Conduct of Parties in the South China Sea (DOC) and the future Code of Conduct in the South China Sea (COC). Fourth, the consensus reached by regional countries through interactions, including the "ASEAN Way" of handling issues, which advocates consensus and accommodates the comfort level of all parties.

China has been a firm defender and active contributor to the international rule of law, norms and orders. As early as in 1954, China, Myanmar and India initiated the Five Principles of Peaceful Coexistence. To conform to the new international order for the seas, China and ASEAN members jointly formulated the DOC in 2002, and have made the best efforts to implement it fully and effectively. This is a great contribution to peace and stability in the South China Sea. In July this year, China and ASEAN members issued a Joint Statement on the Full and Effective Implementation of the DOC. We are steadily pushing forward negotiations and working for early conclusion of the COC on the basis of consensus. We hope countries outside the region could respect and support the endeavors of China and ASEAN countries to maintain peace and stability in the South China Sea.

Third, the security architecture should be held together by partnerships. To a large extent, progress in building the Asia-Pacific security architecture depends on the relationships among regional countries as well as interactions among major countries in the Asia-Pacific. All countries should abandon the Cold War mentality, and work together to pursue a new path of dialogue and partnership, instead of confrontation, alliance, and enmity. We should build an Asia-Pacific partnership featuring equality, mutual trust, inclusiveness and mutual benefit through consultation and mutual accommodation.

China is committed to promoting sound interactions with other major countries, which has contributed to stable relations among major countries in the Asia-Pacific. China is actively pushing forward a new model of major-country relationship with the US based on no conflict, no confrontation, mutual respect and win-win cooperation. The comprehensive strategic partnership of coordination between China and Russia is showing a strong momentum. China is actively implementing partnerships for peace, growth, reform and civilization with Europe. China is actively strengthening the strategic partnership of peace and prosperity with India, and developing a closer partnership for development. And China is also endeavoring to improve its strategic relationship of mutual benefit with Japan based on the spirit of taking history as guidance and looking into the future.

Fourth, the security architecture should be supported by a comprehensive and multi-layered network. Given the diversity in the Asia-Pacific, a unified security framework in this region is not foreseeable in the near future. In fact, the building of the new Asia-Pacific security architecture is not meant as the start of a new venture. Rather, it will be based on the further coordination, upgrading and improvement of existing mechanisms. It will be normal for various mechanisms to advance together and form a multi-layered, wide-ranging and comprehensive Asia-Pacific security framework.

All parties should continue to respect the ASEAN centrality in East Asia cooperation, encourage the SCO and the CICA to play bigger roles, and support the development of Track 1.5 and Track 2 dialogue platforms, such as the Xiangshan Forum. Bilateral military alliances should be more transparent and play a constructive role for regional peace and stability. China will as always support and promote the development of regional security dialogue and cooperation mechanisms, actively participate in them, and improve regional security architecture together with other parties.

Fifth, the security architecture should draw strength from common development. Achieving common development and prosperity is the common aspiration of all countries in the Asia-Pacific. The Asia-Pacific security architecture should also serve this aim. Countries should vigorously promote social and economic development, improve living standards, and narrow the development gap within the region. At the same time, we should pay close attention to sustainable development, and create conditions to enable all countries and people of all social backgrounds to access the fruits of development and achieve inclusive and mutually beneficial development.

To promote common development, China has proposed and actively advanced the Belt and Road Initiative, and initiated the Asian Infrastructure Investment Bank (AIIB) and the Silk Road Fund. To help narrow the development gap within ASEAN countries, China and Mekong countries jointly set up the Lancang-Mekong Cooperation mechanism. With China's efforts, the G20 Hangzhou Summit reached important consensus on inclusive and interconnected development.

Forging Sound Relations with Amity, Sincerity, Mutual Benefit and Inclusiveness to Write New Chapters in Neighborhood Diplomacy

Published on the *People's Daily* on January 10, 2017

In the past year 2016, against the backdrop of changing international situation, Asia, while maintaining overall stability, also faced many factors of instability and uncertainty. How to uphold peace, development and cooperation and maintain the stability and development momentum that has not come by easily in Asia? All eyes are turning to China. In the past year, under the strong leadership of the Communist Party of China (CPC) Central Committee with Comrade Xi Jinping as the core, China's neighborhood diplomacy has overcome many difficulties, made pioneering efforts and forged ahead, vigorously safeguarded national sovereignty, security and development interests, further tightened the bond of the community of shared future with neighboring countries, and spoke loud for unity, cooperation, mutual benefit and win-win results in Asia.

Forge sound relations extensively and move relations with neighboring countries to a new stage

In 2016, the Asian situation was quite complicated with recurring emergencies, which brought new challenges to the relations among China and neighboring countries. We followed the principle of "affinity rather than enmity, partnership rather than alliance", kept firmly in mind the principle of amity, sincerity, mutual benefit and inclusiveness, and put it into practice. In this way, China consolidated good-neighborly friendliness and mutually beneficial cooperation with other countries.

"Neighbors become closer with frequent exchanges." The leading role and demonstration effect of high-level exchanges become more prominent in neighborhood diplomacy. Cambodia has upheld justice and righteousness and maintained friendly relations with China regardless of the external situation. President Xi Jinping paid a state visit to

Cambodia, which deepened China-Cambodia traditional friendship and sent the clear signal that China cherishes and helps old friends. President Xi Jinping's visit to Bangladesh marked the first visit to the country by a Chinese head of state in the last three decades. China became the first strategic cooperative partner of Bangladesh. The two countries reached important consensus on the integration of development strategies and the common development of the Belt and Road Initiative, which set a new model for South-South cooperation. On the occasion of the 55th anniversary of the establishment of diplomatic relations between China and Laos, Premier Li Keqiang visited Laos. The two sides agreed to deepen the comprehensive strategic cooperative partnership and signed a series of important cooperation documents, which consolidated strategic mutual trust and deepened traditional friendship.

In face of the once worsened China-Philippines relations, we pinned our hopes on the Philippine people, kept working on all walks of life in the Philippines and achieved positive outcomes. After President Rodrigo Duterte took office, he actively and properly dealt with the South China Sea issue, improved relations with China and chose China as the destination of his first visit outside ASEAN countries, through which the China-Philippines relationship made a "magnificent turn" and returned to the right track of good-neighborly friendship and healthy development. We grasped the overall picture of China-Vietnam friendship and the development direction of socialism and promoted steady development of China-Vietnam relations after the leadership transition of the Communist Party of Vietnam and the Vietnamese government. We supported Myanmar to follow a development path that is in line with its national conditions and achieve a smooth transfer of political power, and worked with Myanmar to maintain the larger interest of bilateral friendly cooperation. After unremitting efforts, the new Sri Lankan government maintained the general direction of China-Sri Lanka friendship, and bilateral practical cooperation regained the momentum of smooth development. We kept strengthening relations with neighboring countries to expand and deepen our "circle of friends" in Asia.

Actively make plans to build a stable and balanced framework of major-country relations

The year 2016 saw profound changes in the balance of power in

Asia. Whether major countries can get on well with each other directly affects regional peace and stability, We did our best to develop stable and healthy relations with major countries and played a stabilizing role in the regional situation.

We actively built a new type of China-US major-country relations featuring no conflicts, no confrontation, mutual respect and win-win cooperation, carried forward the candid exchanges and strategic communication between both heads of state, expanded common interests and managed and controlled differences and contradictions to build an Asia-Pacific interactive framework centered on China-US cooperation. We strengthened strategic coordination with Russia on international and regional affairs, worked together to cope with all kinds of risks and challenges, and maintained regional peace, stability as well as strategic balance. We amplified our role in leading and shaping and maintained the overall improving momentum of China-Japan relations. Bilateral exchanges in various sectors resumed in an orderly manner. We firmly grasped the general development direction of China-India relations. Leaders of the two countries met with each other several times this year, steadily pushing forward the ever-closer partnership of development.

Manage and control hotspot issues to maintain the two major interests of peace and development

The year 2016 was not a peaceful year for China's surrounding situation as old and new hotspot issues kept emerging, adversely affecting peace and stability in Asia. We never tolerate any action that undermines China's sovereign rights and interests and we never hesitate in the cause of maintaining regional peace and stability.

The greatest battle for us was dealing with the South China Sea arbitration case. A handful of regional and non-regional forces used the South China Sea arbitration case as a cover to pursue their hidden agenda of infringing upon China's interests and mounting an all-out demonizing campaign against China. We reacted to this issue by arguing strongly on just grounds and fighting back resolutely, during which we gained the understanding and support from nearly 120 countries and about 240 political parties in different countries. We actively engaged with ASEAN countries, jointly issued the Joint Statement on the Full and Effective Implementation of the Declaration on the Conduct of Parties in the South China Sea and applied the dual-track approach to guide each party to focus on dialogue and cooperation. We stepped forward to turn around the China-Philippine relations and guided the South China Sea issue back on the track of bilateral dialogue and negotiation, realizing a "soft landing" of the arbitration case. China's stance is reasonable and legitimate, and our actions are above board. We brought back the prospect of peace, stability and cooperation in the South China Sea, and upheld, in a true sense, the order and dignity of the international law.

In light of the tense situation on the Korean Peninsula, we remain committed to denuclearization of the Korean Peninsula, peace and stability on the Peninsula and resolution of the issue through dialogue and negotiation. We proposed a dual-track approach in which the denuclearization of the Korean Peninsula and transformation from the armistice regime to a state of peace proceed together, and made great efforts to pull the nuclear issue back to the negotiation table. Resolute in safeguarding our legitimate interests and strategic security environment, we opposed the deployment of the THAAD missile defense system on the Peninsula under the pretext of the nuclear issue. Cherishing the China-Myanmar "paukphaw" friendship and based on the will of Myanmar, we played a constructive role in the peace and national reconciliation process of Myanmar. We actively took part in the Istanbul Process on Afghanistan by providing strong support to the domestic political reconciliation in Afghanistan.

Focus on cooperation and promote the Belt and Road development both in substance and depth

In 2016, the world economy went through sluggish recovery. Asian countries faced mounting downward economic pressure with salient structural problems. We concentrated on the Belt and Road development and docked development strategies, advanced connectivity and deepened cooperation in production capacity with neighboring countries to inject strong impetus into Asia's development.

We made steady headway in infrastructure connectivity. The China-Pakistan Economic Corridor has entered into the stage of comprehensive implementation and striving for early harvest, becoming the project proceeding most rapidly in the comprehensive development of the Belt and Road Initiative outside China. We formulated the outline for

China-Mongolia-Russia Economic Corridor and initiated the inter-governmental cooperation process for the Bangladesh-China-India-Myanmar Economic Corridor. We held the launching ceremony of the China-Laos Railway project, signed the framework agreement on inter-governmental cooperation for China-Thailand Railway and began to build the guiding section of the Jakarta-Bandung High-speed Railway. China and Malaysia conducted cooperation on the Malaysia Southern Railway project. A Chinese enterprise won the bid of Myanmar's Kyaukpyu deep water port and industrial zone project. The second phase of the Hambantota port in Sri Lanka advanced orderly and the Colombo port city project resumed comprehensive construction.

Production capacity cooperation was carried out comprehensively. China signed joint declarations of production capacity cooperation successively with ASEAN countries and member countries of the Lancang-Mekong Cooperation mechanism, and accelerated the development of industrial zones, cross-border economic cooperation zones and harbor industrial park projects with Southeast Asian and South Asian countries. A batch of early-harvest projects has yielded initial benefits. Innovation in the Belt and Road mechanism made new progress. The Asian Infrastructure Investment Bank has been put into operation and the first batch of projects has been successfully launched. The first investment projects of the Silk Road Fund were also officially initiated.

Strengthen guidance to consolidate and expand regional cooperation momentum

The year 2016 witnessed setbacks in economic globalization and regional integration. Regional cooperation in East Asia suffered from increasing obstacles and difficulties. Holding high the banner of openness, inclusiveness and win-win cooperation, China played an active leading role and took active actions to lead the development of regional cooperation architecture.

Premier Li Keqiang, together with leaders from ASEAN members, attended the Commemorative Summit Marking the 25th Anniversary of China-ASEAN Dialogue Relations. They all agreed to strengthen strategic communication, deepen practical cooperation and promote the implementation of the Belt and Road Initiative, the "2+7 cooperation framework" and other important initiatives. China and ASEAN member states held a series of events to celebrate the 25th anniversary of China-ASEAN dialogue relations and the China-ASEAN Year of Education Exchange. China-ASEAN relations clearly showed a positive momentum and accelerated the transition from the "growth stage" of rapid development for the past 25 years into the "mature stage" of quality improvement in the upcoming 25 years. China has successfully held the first Lancang-Mekong Cooperation Leaders' Meeting, formulated rules and regulations for the Lancang-Mekong Cooperation, identified the "3+5 cooperation framework" and 45 early-harvest projects, actively led the Lancang-Mekong Cooperation to enter a new phase of comprehensive implementation and put forward China's proposal to benefit people of countries along the Lancang-Mekong River. The Lancang-Mekong Cooperation mechanism has become another important platform for China and neighboring countries to build a community of shared future.

We firmly seized the right direction of regional cooperation in East Asia, maintained the development momentum of China-Japan-ROK cooperation, "10+3" (ASEAN plus China, Japan and ROK) cooperation and other cooperation mechanisms and guided the East Asia Summit (EAS) to get rid of disturbing factors and focus on the "greatest common divisor" between economic development and political security cooperation. We actively practiced the common, comprehensive, cooperative and sustainable security concept for Asia and promoted practical cooperation in non-traditional security fields and discussions on building a new regional security framework based on regional realities. We continued to advance the negotiation process of China-Japan-ROK FTA and Regional Comprehensive Economic Partnership (RCEP). And we made unremitting efforts to achieve the goal of the FTA of the Asia-Pacific (FTAAP). We have successfully held the Boao Forum for Asia Annual Conference to make China heard and spread opportunities for Asia, enhancing the confidence of all sides in the development prospect of China and Asia. We actively participated in the process of the Asia Cooperation Dialogue (ACD), devoting ourselves to promoting pan-Asia cooperation.

In 2017, we will embrace the 19th CPC National Congress and the 13th Five-Year Plan will enter a critical year of comprehensive and in-depth implementation. The year 2017 is also a year for the international system to experience significant transformation and profound changes.

Throughout the world, the theme of peace and development remains unchanged, yet all kinds of turbulences still exist, posing greater risks and challenges to the neighborhood diplomacy. Facing the changing and unstable international situation, all countries will pay more attention to Asia and expect more from China. We will coordinate the domestic and the international situation as well as the two major interests of development and security, and seek progress in stability to make greater progress in neighborhood diplomacy.

We will focus on neighboring countries to carry out high-level exchanges and strategic communication. We will also take such home-field diplomacy as the International Forum on Belt and Road cooperation, the ninth BRICS Summit and the Boao Forum for Asia Annual Conference as opportunities to make steady headway in the Belt and Road development in neighboring countries, strengthen the bonds of interests with neighboring countries, and stimulate the endogenous dynamics of regional cooperation. While safeguarding China's sovereignty and maritime rights and interests, China will also properly handle disputes and differences through dialogues and negotiations. China will continue to promote peace and facilitate discussions on the nuclear issue on the Korean Peninsula, the South China Sea issue, the Afghan issue and other issues to maintain regional peace and stability. We should vigorously boost regional cooperation for in-depth and substantial results, and consolidate the FTAAP system with East Asia as its center. We should also carry forward the spirit of Asian civilization and reinforce people-to-people and cultural exchanges and inter-civilization dialogues to enhance regional identity and friendly sentiments among the people.

As a new year begins, everything will be fresh and new. In the new year, China's neighborhood diplomacy will continue to serve the cause of national security and reform and development to create a sound environment for the successful convening of the 19th CPC National Congress and the realization of the first of the "two centenary goals". We will regard Asia's peace, stability, prosperity and development as our own duty, help and benefit our neighboring countries, and write a new chapter in the building of the community of shared future.

Steadily Advancing Asia-Pacific Regional Security Cooperation

Excerpts from the press conference on China's Policies on Asia-Pacific Security Cooperation
January 11, 2017

The world is undergoing complex and profound changes both politically and economically. Globalization has met severe headwinds and the integration processes of some regions have hit setbacks. As the world goes through a period of volatility, the Asia-Pacific, thanks to the deepening cooperation among regional countries, has maintained its momentum of rise on the whole and assumed a more prominent role in the world. Long committed to peace and security in the Asia-Pacific, China issued a White Paper on China's Policies on Asia-Pacific Security Cooperation at the beginning of the year to expound on its policy on Asia-Pacific security cooperation and demonstrate its readiness to further strengthen security cooperation and uphold stability and prosperity in the region. We hope that the white paper will receive positive response from across the region. We look forward to working with countries in the region to steadily advance security cooperation and make greater contribution to peace and stability in the region.

For a better understanding of the white paper, let me use six key phrases to illustrate China's policy on Asia-Pacific security cooperation.

The first phrase is win-win cooperation. To maintain the region's strong development momentum and pool efforts to tackle the challenges, China made a six-point proposal on strengthening Asia-Pacific security dialogue and cooperation, namely promoting common development, building partnerships, improving the existing regional multilateral mechanisms, advancing rule-making, intensifying military exchanges and cooperation, and properly resolving differences and disputes. This proposal draws its elements from the experience China has gained from its long history of engagement with other regional countries and the shared wisdom of countries in this region. It is hoped that concerted efforts will be made in pursuing win-win cooperation and enduring peace and security in this region.

The second phrase is openness and innovation. China believes that the regional vision and framework for security must advance with time and be open to innovative ideas to meet the region's security needs and resolve specific issues. In this context, China put forward the vision of common, comprehensive, cooperative and sustainable security and envisioned a future-oriented security framework which is geared towards the region's realities and accommodates the security needs of all parties. This security vision, grounded in the region's tradition, aims to introduce new ideas about cooperation and inject fresh vitality to the development of the Asia-Pacific security framework. China hopes to work with all parties to advocate and practise the new security vision and pursue a shared security path for the Asia-Pacific that benefits all.

The third phrase is sound interactions. China engages in sound interactions with other major countries in the Asia-Pacific, contributing to the overall stability of major-country relations in the region. In its relations with the United States, China is committed to building a relationship featuring non-conflict, non-confrontation, mutual respect and win-win cooperation, with sound interactions and inclusive collaboration in the Asia-Pacific. In its relations with Russia, China is committed to strengthening the comprehensive strategic partnership of coordination. We are committed to fostering a closer partnership for development with India, and work to improve and further develop our relations with Japan in the spirit of taking history as a mirror and looking to the future. We will continue to build friendship and partnership with our neighbors, deepening mutually beneficial cooperation.

The fourth phrase is dialogue and consultation. What has happened since World War II shows that dialogue and cooperation are the keys to resolving regional hotspot issues. We believe that disputes and differences among countries should be properly settled through dialogue and negotiation to jointly uphold peace and stability in the region. The cross-border security challenges facing the entire region call for joint response through more dialogue and cooperation, the only way to prevent the challenges from turning into crises. China will continue to play its role as a responsible major country in seeking peaceful settlement of regional hotspot issues.

The fifth phrase is regional mechanism. As a strong advocate of and contributor to regional multilateral mechanisms, China is a founding member of many such mechanisms. We firmly support ASEAN's centrality in East Asian cooperation and have been actively involved in and promoted dialogue and cooperation in relevant mechanisms. China is committed to advancing the development of regional mechanisms and improving regional cooperation frameworks. By working with relevant countries, we have established the Shanghai Cooperation Organization, launched the Six-Party Talks on the Korean nuclear issue, the Xiangshan Forum and the mechanism for Lancang-Mekong Cooperation to advance economic cooperation as well as political and security cooperation, the two pillars of regional cooperation. China will continue to support and promote even greater development of regional cooperation mechanisms.

The sixth phrase is results-oriented cooperation. In light of the growing non-traditional security threats in recent years, China has taken an active part in exchanges and cooperation on tackling non-traditional security challenges. It has provided security-related public goods to other countries and carried out a large number of cooperation projects on disaster relief, counter-terrorism, combating transnational crime, cyber security and maritime security. These efforts have deepened the mutual understanding and trust among countries in the region, and enhanced their capabilities in making coordinated responses to challenges, contributing to regional peace and stability.

Several transregional security cooperation mechanisms exist along with each other in the Asia-Pacific region, including the US-dominated alliances system, the ASEAN-centered security dialogue and cooperation framework, and specialized mechanisms such as the Six-Party Talks, as well as the Shanghai Cooperation Organization. Some are left by the Cold War, while some others are the result of lack of coordination among sub-regions and differences on security concepts.

Our region's security cooperation architecture lags behind its relatively mature and stable economic cooperation framework. The architecture is unable to respond promptly and effectively to the complex and diverse challenges the region faces. It is a vital task to put in place an Asia-Pacific security cooperation architecture which is future-oriented, consistent with regional conditions and in the interests of all parties.

The new security architecture for the Asia-Pacific must have the following five features: Firstly, it must be guided by the new vision of common, comprehensive, cooperative and sustainable security;

Secondly, it must be based on international rules and norms and consistent with the purposes and principles of the UN Charter and rules governing regional and international relations agreed on by all countries; Thirdly, it should be held together by partnerships, in particular an Asia-Pacific partnership of equality, mutual trust, inclusiveness, mutual benefit, mutual understanding and mutual accommodation; Fourthly, it should be supported by a diverse, comprehensive and multi-layered network based on further coordination, upgrading and improvement of existing mechanisms; and Fifthly, it should draw strength from common development that is inclusive and mutually beneficial.

Before a new regional security architecture is put in place, countries in the region need to intensify track 1.5 and track 2 research and consultation.

To conclude, as a key member of the Asia-Pacific community, China has a keen sense of responsibility to advance the prosperity, stability and security of the Asia-Pacific. We will work with other countries in the region to enhance security dialogue and cooperation, and tackle traditional and non-traditional security challenges, with a view to jointly advancing peace, stability and common prosperity in the Asia-Pacific.

Pushing China-ASEAN Relations to a New Stage

Speech at the meeting with ASEAN Ambassadors to China
February 10, 2017

In 2016, economic globalization and regional integration suffered setbacks. Regional cooperation in East Asia encountered increasing resistance and difficulties. And non-regional powers' intervention in the South China Sea issue added destabilizing factors to the region and disrupted the East Asian cooperation process. With joint efforts by China and ASEAN countries, we successfully celebrated the 25th anniversary of the establishment of China-ASEAN dialogue relations, overcame the difficulties and challenges brought by the South China Sea issue, effectively advanced China-ASEAN relations, and made great contributions to keeping the momentum of East Asian cooperation. It is fair to say that China-ASEAN relations are moving from a phase of rapid growth to one of higher quality development.

This year marks the beginning of another 25-year period for China-ASEAN relations and the 50th anniversary of ASEAN. When we look back, in the first half of the past five decades, Southeast Asia suffered from turbulence and division due to the Cold War. During the second 25 years, after the end of the Cold War and the peaceful settlement of the Cambodian issue, Southeast Asian countries enjoyed a golden period of development and ASEAN grew stronger with its members increasing to ten. While building its own capacity, ASEAN also forged dialogue partnerships with China and over ten other countries and regional and international organizations. The ASEAN Communities were established in 2015. We are happy to see ASEAN's progress and look forward to a more united, stronger and more prosperous ASEAN in the future.

ASEAN and China-ASEAN relations face new tasks and challenges under the new circumstances. Given the trend of de-globalization, ASEAN is confronted with new tests. Internally, it needs to narrow the development gap among member countries and upgrade cooperation in the Community. Externally, it needs to balance and properly handle the relations with its dialogue partners and consolidate its centrality in East Asian cooperation. China and ASEAN countries need to deepen mutual trust, explore internal dynamism, and elevate cooperation to a new level.

Not long ago, President Xi Jinping delivered a keynote speech at the World Economic Forum Annual Meeting at Davos, demonstrating China's positive stance on and firm support for economic globalization. In our view, the progress of ASEAN, a model for regional integration, provides strong support for globalization. We hope that in the next 25 years, ASEAN will accelerate the development of the ASEAN Community with redoubled efforts, and focus on cooperation and development, promote regional peace and stability, and contribute more to the agreed goals of building the East Asia Economic Community and a community with a shared future for Asia.

As President Xi Jinping pointed out in his 2017 New Year message that no pie will fall from the sky and we must roll up our sleeves to work harder. This line, full of positive energy, has been applauded by the Chinese people and frequently quoted online. Great visions can only be realized through pragmatic actions. In growing China-ASEAN relations, we also need to work hard to deepen cooperation in seven priority areas and foster new growth points, focusing on the three pillars of political and security issues, economic and sustainable development, and social, cultural and people-to-people exchanges. In this connection, the Chinese side suggests that we focus on the following six areas in the next phase:

First, make good preparations for the meetings for East Asian cooperation this year. The foreign ministers' meetings in August and the leaders' meetings in November will be important opportunities and platforms for us to upgrade China-ASEAN cooperation. We need to make thoughtful preparations to ensure successful and productive meetings. To this end, we suggest that the 23rd China-ASEAN Senior Officials' Consultation be held in China as soon as possible at a time convenient to all parties back to back with the Senior Officials' Meeting on the Implementation of the Declaration on the Conduct of Parties in the South China Sea in preparation for the leaders' meetings and the foreign ministers' meetings.

Second, foster greater synergy between development strategies. China will seek greater complementarity between the Belt and Road Initiative and the development strategies of ASEAN, particularly the

Master Plan on ASEAN Connectivity 2025, promote the delivery of more cooperation projects, and strengthen connectivity cooperation in a vigorous and orderly manner. We believe that our joint efforts will inject strong impetus into regional economic growth and the momentum of East Asian cooperation.

Third, facilitate the work on the pillar of social, cultural and people-to-people exchanges. Both sides need to take the China-ASEAN Tourism Cooperation Year as an opportunity to vigorously expand tourism exchanges and establish a tourism cooperation mechanism. China is considering holding the inauguration ceremony of the Tourism Cooperation Year with ASEAN in the Philippines in mid-to-late March. Our two sides are keeping in close contact with each other. Both sides need to strengthen educational cooperation and ensure the success of the celebrations for the tenth anniversary of China-ASEAN Education Cooperation Week. China is ready to provide concrete support to the commemoration for the 50th anniversary of ASEAN.

Fourth, further deepen pragmatic cooperation. We should implement the Plan of Action to Implement the Joint Declaration on ASEAN-China Strategic Partnership for Peace and Prosperity (2016-2020), follow through on the existing initiatives, and promote the early delivery of the outcomes of the upgraded China-ASEAN FTA. China will deepen cooperation with ASEAN in such areas as agriculture and poverty alleviation and scale up cooperation in sustainable development.

Fifth, create new growth drivers for sub-regional cooperation. Lancang-Mekong Cooperation (LMC) has entered a new stage of implementing cooperation programs and has delivered a series of pragmatic results. China will assist ASEAN in narrowing development gaps among member states through the LMC and welcomes support and engagement from other ASEAN countries. China will play an active role in sub-regional cooperation initiatives such as the East ASEAN Growth Area, and create new engines for East Asian cooperation so as to contribute to the development of the ASEAN Community. China and ASEAN may start from some concrete projects to explore win-win cooperation.

Sixth, properly handle the South China Sea issue. After turbulence and adjustment last year, the issue has been brought back to the track of dialogue and consultation. One important lesson is that intervention by non-regional forces cannot bring real peace and tranquility to the South

China Sea. It will only create new problems for regional stability and undermines the interests of China and ASEAN countries. Our priority this year is to continue the comprehensive and effective implementation of the DOC and produce a COC draft framework, free from outside interference, as soon as possible in the first half of this year.

China's position on the South China Sea is consistent. China's primary interest is peace and stability of the South China Sea, which is in the common interests of China and ASEAN countries. I believe that ASEAN countries know full well that peace and stability in the South China Sea bears on security and development in the region. China and ASEAN countries need to work together to keep the momentum of the "dual-track approach" to properly manage the South China Sea issue.

Taking ASEAN as a priority in its neighborhood diplomacy, China firmly supports the development of the ASEAN Community, and ASEAN's centrality in regional cooperation and a greater role in regional and international affairs. Given the many uncertainties in East Asian cooperation, we hope that ASEAN will continue to uphold the principles of maintaining its centrality, building consensus, accommodating each other's comfort level, and serve as a good "designer" and "driver" for regional cooperation. We will provide support and assistance for ASEAN to play its role.

On ASEAN Plus Three (APT) cooperation, as we celebrate its 20th anniversary this year and take it as a new starting point, all parties need to promote cooperation in such priority areas as trade and finance, food security and poverty reduction, and consolidate APT's role as the main channel of regional economic integration. To achieve the goal of an East Asia Economic Community (EAEC), China calls for drafting an EAEC Blueprint to identify the specific indicators and means for building the community. We count on the active support from ASEAN in this regard. China will host the 15th East Asia Forum this year and launch commemorative activities for the 20th anniversary together with the forum. We look forward to active participation by ASEAN. China will work with ASEAN to accelerate the negotiations on the Regional Comprehensive Economic Partnership (RCEP), and work for the conclusion of the negotiations within this year to demonstrate the determination of East Asian countries to promote trade liberalization. China supports ASEAN in its continued leadership role in RCEP negotiations.

On the East Asia Summit (EAS), we need to adhere to the positioning of the EAS as a "leaders-led strategic forum", pursue economic and social development and political and security cooperation as the two drivers, and avoid hyping up political and security issues, so that the EAS will play a greater role in maintaining peace and stability and boosting development and prosperity in the region. All parties need to focus on development and cooperation, implement well the Plan of Action to Implement the Phnom Penh Declaration and carry out more cooperation programs.

China will work with other parties to step up exchanges on regional security concept and framework with a view to establishing a regional security architecture that is suited to the region's reality and meets the needs of various parties. Thailand will hold the sixth Workshop on Regional Security Architecture in the middle of this year. And China will hold the Track 2 Seminar on Regional Security Architecture. We welcome active participation by ASEAN countries.

On the ASEAN Regional Forum (ARF), we believe that the ARFs need to continue to focus on enhancing mutual trust, prioritize non-traditional security cooperation, take sound interactions among major countries as a key task, and explore, step by step, preventive diplomacy that suits conditions in this region. China has played an active part in pushing forward pragmatic cooperation within the ARF framework. China ranks at the top in the number of practical cooperation projects it hosts, which accounts for one third of all ARF projects every year in recent years. We will further enhance cooperation with ASEAN countries and launch more high-quality cooperation projects on non-traditional security.

The White Paper on China's Policies on Asia-Pacific Security Cooperation issued last month fully demonstrates China's focus and positive stance on regional security cooperation and its resolution to pursue peaceful development and maintain regional stability and prosperity. We stand ready to work with ASEAN to deepen regional security cooperation and make even greater contribution to regional peace, stability and prosperity.

Steadily Promoting Cooperation among South China Sea Coastal States

Speech at the South China Sea Session of Boao Forum for Asia Annual Conference 2017
Boao, Hainan, March 25, 2017

Over the last 20 plus years since the end of the Cold War, Southeast Asian countries have enjoyed the longest period of peace in the post-World War II era, marked by fast growth and rising living standards. In the spirit of unity and mutual help, China and ASEAN countries carried out fruitful cooperation in various fields. Last year, we celebrated the 25th anniversary of China-ASEAN dialogue relations. China has become the largest trading partner of ASEAN, and we look forward to working with ASEAN to embrace the next 25 years of our booming ties.

With the joint efforts of China and ASEAN countries, the situation in the South China Sea has been stable overall and moving in a positive direction. With the personal attention of state leaders, China and the Philippines reached important consensus on properly handling relevant issues in the South China Sea, bringing about a turnaround in bilateral relations. This January, the two sides agreed to set up a bilateral consultation mechanism on the South China Sea issue, and the first meeting will soon be held. Furthermore, China and ASEAN countries issued the Joint Statement on the Full and Effective Implementation of the Declaration on the Conduct of Parties in the South China Sea (DOC) last July and other important documents, reaffirming that the relevant disputes should be settled through negotiation and consultation by countries directly concerned. It sends a strong signal that regional countries will work together to maintain peace and stability in the South China Sea. In my visits to Southeast Asian countries since January, I shared my thoughts on promoting cooperation among South China Sea coastal states. ASEAN colleagues showed great interest in and wanted to further explore this idea. Today, let me focus my remarks on this topic.

The South China Sea is a typical closed or semi-closed sea. Peace and stability there is crucial for the security, development and prosperity of all coastal states and the well-beings of their peoples. It is in the shared interests of all coastal states to promote peace, stability, prosperity and development, which is also our common responsibility. Now is the right time to launch cooperation among South China Sea coastal states. To quote a Chinese idiom, it is "of the right time, in the right place and with the support of the people." The coastal states should address the South China Sea issue with an open mind and creative approach, and stay focused on cooperation while properly handling differences. They should carry forward traditional friendship, maintain peace and stability in the South China Sea, which is conducive to cooperation, development and prosperity in the whole region.

As I just said, now is the right time to launch cooperation among coastal states. The situation in the South China Sea has been cooling down. With countries returning to the right track of settling disputes through negotiation and consultation, much can be achieved in practical maritime cooperation. As early as six years ago, the Chinese government declared the establishment of the China-ASEAN Maritime Cooperation Fund, and has begun to finance relevant cooperation programs. In 2013, President Xi Jinping put forward the important initiative of building a 21st Century Maritime Silk Road during his visit to Southeast Asia. Now, the Asian Infrastructure Investment Bank (AIIB) and the Silk Road Fund are ready to support relevant cooperation projects. As of today, more than 100 countries and international organizations have endorsed China's Belt and Road Initiative, and over 40 of them have signed cooperation agreements with China. Chinese companies have invested over US$50 billion in countries along the Maritime Silk Road and implemented a number of major projects, making a positive impact on the economic growth of relevant countries.

China will continue to align its development strategy with those of ASEAN countries, especially the Master Plan on ASEAN Connectivity 2025 (MPAC2025). We will strengthen cooperation on production capacity and implement more cooperation projects in ASEAN countries, to lend new impetus to growth and regional cooperation in East Asia. Countries in the region are welcome to ride on the express train of China's development.

It is the right place to launch cooperation among coastal states in the South China Sea. There is solid international legal ground for setting up such a cooperation framework. Similar cases of cooperation can be found around the world. According to the 1982 United Nations Convention of the Law of the Sea (UNCLOS), states bordering an enclosed or semi-enclosed sea should cooperate with each other in the exercise of their rights and in the performance of their duties under UNCLOS. They shall endeavor, directly or through an appropriate regional organization, to coordinate the utilization of the living resources of the sea, and protect and preserve the marine environment. Moreover, many international organizations, including specialized agencies of the United Nations, have encouraged the coastal states bordering an enclosed or semi-enclosed sea to establish mechanisms of cooperation.

Since the 1950s, the coastal states of many closed or semi-closed seas, such as the Caribbean Sea, the Mediterranean Sea, the Baltic Sea, the Black Sea and the Caspian Sea, have established cooperation mechanisms. Some are mature cooperation systems with multiple frameworks and institutions, and some have focused on thematic cooperation such as maritime security, search and rescue, marine environmental protection, fishery management and marine scientific research. These mechanisms broadened and enriched regional cooperation. They have been helpful in enhancing mutual trust among coastal states and mitigating territorial and jurisdictional disputes. These are valuable experience for cooperation among the South China Sea coastal states.

In fact, the South China Sea coastal states have made some successful attempts in this direction. We carried out some cooperation programs on maritime search and rescue, marine scientific research and environmental protection within the framework of the full and effective implementation of the DOC. At the non-government level, we conducted a number of cooperation projects concerning people's livelihood through the Workshops on Managing Potential Conflict in the South China Sea. All parties welcome and wish to continue such cooperation. With the endorsement from Dr. Jalal, many cooperation projects have been carried out.

Advancing cooperation among coastal states is supported by the people. People in our region enjoy a time-honored friendship and formed close cultural bonds. And there is a long tradition of cooperation among coastal states. Since ancient times, the South China Sea has been a major channel for trade and people-to-people exchanges among China and countries in Southeast Asia, South Asia, West Asia and even Europe and Africa. As a vital part of the Maritime Silk Road, the South China Sea played an important part in strengthening political relations, trade and friendship among China and foreign countries. It has borne witness to mutual help and common progress of all coastal states.

As a major channel on the ancient Maritime Silk Road, the South China Sea carries the beautiful memories of the coastal people weathering ups and downs and making progress together. In the context of profound friendship among coastal states, the disputes in the South China Sea, which surfaced a few decades ago, only account for a small fraction of our relations. I see no reason that we cannot do a better job than our ancestors.

The South China Sea coastal states should draw upon the successful experience of other regions, and without prejudice to the parties' respective claims, work together to initiate and establish a relevant cooperation framework, which could serve as an effective platform for enhancing mutual trust, strengthening cooperation and promoting shared interests. Such a framework can bring the relevant states together in practical and institutionalized cooperation in fields such as disaster prevention and reduction, maritime search and rescue, protection of marine environment and bio-diversity, marine scientific research and navigation safety. We may also exchange views on relevant maritime issues in a proper way. We are confident that the establishment of such a framework will be conducive to mutual trust and common development among regional countries.

In addition, I want to stress the following two points:

First, such a framework will run in parallel and be complementary to the existing bilateral cooperation mechanisms among China and ASEAN countries and the multilateral consultation mechanism within the DOC framework. It will only provide a new platform for relevant technical cooperation.

Second, the purpose of this framework is to enhance practical cooperation and mutual trust, it is not concerned with dispute settlement. Cooperation among the coastal states will not hamper the efforts of sovereign states directly concerned to seek peaceful settlement of territorial and jurisdictional disputes through negotiation and consultation, nor will it prejudice the parties' claim and position on the relevant issue.

I have conveyed the above ideas informally to colleagues from South China Sea coastal states and they have responded positively. We are ready to engage in more communication and coordination with parties concerned. We welcome responses and suggestions from regional countries.

Toward a Community with a Shared Future for Island Economies

Speech at the "21st-Century Maritime Silk Road: Islands Economic Cooperation Sub-Forum" of the Boao Forum for Asia Annual Conference
March 25, 2017

The ocean is a common home for humankind and an important channel for the flow of economic resources and wealth. The marine economy in various countries has developed rapidly in recent years. In some East Asian countries, the marine economy accounts for close to 20 percent of their GDP. In 2016, China's gross marine product exceeded RMB7 trillion yuan, nearly one tenth of its GDP. The marine economy is increasingly becoming a new growth area of economic development and regional cooperation.

Island states survive and thrive by the sea. Building maritime corridors, developing the marine economy and utilizing marine resources are the common pursuits of island economies. Under the current circumstances, I believe that island economies can seize and make good use of two major opportunities for development:

First, the opportunity from the rebalance of the global economy. The new round of scientific and technological revolution is on the horizon, and the reshaping of the international market and division of labor is accelerating. Changes in people's lifestyles and consumption patterns have nurtured new industries and new forms of businesses, including coastal tourism, mariculture, marine environmental protection, and Internet of Things. All parties need to seize the momentum, tap into their natural resources and ecological environment, develop the circular economy and green industries, and build demonstration zones of sound marine ecology. In this way, we will achieve harmony between man and nature, and between human and the ocean as we build beautiful islands with both modern civilization and idyllic scenery.

Second, the opportunity from the Belt and Road Initiative (BRI). Over the past three years since the BRI was proposed, relevant cooperation has been advanced steadily, attracting support and active participation by more than 100 countries and international organizations. From the Pacific Ocean to the India Ocean, from the South China Sea to the Mediterranean Sea, the BRI has formed new "circles of friends" at sea. In May this year, China will host the Belt and Road Forum for International Cooperation in Beijing for joint discussion of cooperation plans, development of cooperation platforms and sharing the outcomes of cooperation so that the BRI will deliver greater benefits to people of all partner countries. Parties are encouraged to make use of this platform to seek synergy between their development strategies and industries, deepen all-round cooperation, explore the potential of economic growth, and expand the space for marine development.

As a proverb goes, "If you want to go quickly, go alone. If you want to go far, go together." Win-win cooperation is the key for island economies to enhance economic strength and international competitiveness. I believe that China is well-positioned to cooperate with island economies in the following three areas:

First, we may work together to maintain peace and security at sea. We need to further improve bilateral and multilateral mechanisms, protect freedom of navigation and safety of passages at sea, and jointly respond to marine disasters in order to maintain a peaceful and tranquil marine order and environment. Second, we may deepen cooperation on connectivity. We may promote aviation and shipping cooperation and facilitate the construction of key seaports to build smooth, efficient, and safe port connectivity networks and reduce the logistics costs, so as to promote sustainable development with the advantages of locations and resources. Third, we may increase cooperation in marine industries. China will actively support island economies in economic diversification and industrialization. Joint efforts will be made to develop modern agriculture, tourism, and service industries, accelerate the exploration of offshore financial and free trade zones, develop complete and diverse marine industrial and value chains to provide more marine public goods.

Deepen Regional Cooperation in Asia with Renewed Confidence

Speech at the "Roundtable on Asian Regional Cooperation Organizations" of the Boao Forum for Asia Annual Conference 2017
March 26, 2017

Over the past year, given the sluggish global growth and the backlash against globalization, the process of globalization and regional integration suffered setbacks, and international political and economic uncertainties have been on the rise. In contrast, regional cooperation in Asia has continued to enjoy a strong momentum, becoming a star-performer in global development.

First, regional cooperation frameworks are booming. East Asia cooperation, with ASEAN at the center, is moving forward with strong vitality. As ASEAN celebrates its 50th anniversary, the building of ASEAN Community is stepping to a higher level. AMRO has been upgraded to an international organization. China-ASEAN cooperation has moved to a new level. And China-Japan-ROK cooperation has made new headway. Pan-Asia cooperation has gained new impetus, with institution building of the Asia Cooperation Dialogue summit meeting. And under CICA, confidence-building measures in various fields have been implemented with positive outcomes.

Second, sub-regional cooperation is gathering momentum. The Lancang-Mekong Cooperation has got off to a good start. The BCIM economic corridor is advancing steadily. The Shanghai Cooperation Organization successfully realized enlargement. BIMP-EAGA has made solid progress. And SAARC cooperation continues to move forward.

Third, fruitful results have been achieved in cooperation across the board. Smooth progress has been made in linking up the Belt and Road Initiative with national plans of relevant countries. Regional connectivity has been strengthened with the construction of major infrastructure projects, including the China-Pakistan Economic Corridor, Jakarta-Bandung High-speed Rail and China-Laos Railway. Regional countries are dedicated to free trade. Negotiations on RCEP and China-Japan-ROK Free Trade Area are moving forward at a faster pace. Regional security dialogue and cooperation are deepening. Countries have acted with greater commitment to jointly combat non-traditional security threats, and conducted several exercises on disaster relief, counter-terrorism and maritime search and rescue. Some have carried out joint law-enforcement operations against drug trafficking and telecommunications fraud.

Looking across Asia, a multi-tier, multi-pillar, all-dimensional network of cooperation has been emerging. Mutual complementarity among various frameworks, deeper cooperation in various fields, and common progress of sub-regional cooperation institutions have become defining features of regional cooperation in Asia.

Such cooperation has given a strong boost to regional integration and played an important part in peace, stability, development and prosperity of the region. It has opened bright prospects for a region-wide cooperation framework and the building of an Asian community of shared future, injecting strong, positive energy into economic globalization.

With international and regional landscape undergoing profound changes, Asian countries have come to a new starting point and are faced with new opportunities for regional cooperation. Faced with sluggish recovery, rising protectionism and growing transnational threats, they have further realized the importance of harnessing complementary advantages for win-win cooperation. It is clearer than ever that closer cooperation is what Asian countries truly need. Last year, regional countries made joint efforts to cool down hotspot issues, and brought relevant issues back to the right track of resolution despite outside interference. That has enabled various sides to re-focus on cooperation for common development. With TPP facing an uncertain future, various sides are shifting their attention to RCEP and FTAAP, showing a greater interest in reaching these FTAs at an early date.

Meanwhile, cooperation in Asia is also confronted with multiple challenges. Internally, regional cooperation has entered a plateau. Many agreed arrangements have not been implemented effectively. This requires countries to make necessary policy adjustments and show greater political wisdom and resolution. In a wider context, the tide of anti-globalization may trigger worries about the future of Asian integration. Some countries outside the region continue to play up political and security issues, causing disturbance to regional cooperation.

China is an active player and advocate of regional cooperation in Asia. In recent years, China has made important contribution to regional cooperation by seeking to enhance its breadth and depth.

First, we supported the development of regional cooperation frameworks. China and the five countries along the Mekong River have launched the Lancang-Mekong Cooperation, which is a useful complement to the ASEAN Community and a worthy effort in building an Asian community of shared future. China has shown strong support for the 50th anniversary of ASEAN by taking an active part in the commemorative events. As the CICA Chair from 2014 to 2018, China has sought creative ways to improve the CICA platform. It has hosted a non-governmental forum and a think tank roundtable. China has proposed a systematic plan for strengthening the ACD, which has been commended by all parties.

Second, we contributed to the cultivation of regional cooperation philosophy. At the 2015 Boao Forum conference, President Xi Jinping outlined his vision for a community of shared future. President Xi has since put forth the concepts of a community of shared future between China and ASEAN, in the Asia-Pacific and, more broadly, for all mankind. They have resonated well with many countries. President Xi has also proposed a new type of international relations underpinned by win-win cooperation. These new ideas have pointed a way forward for regional cooperation in Asia.

Third, we have actively advanced practical cooperation. China champions the Belt and Road Initiative for shared development of the countries along the routes. The AIIB has started operation with the first batch of its projects approved. The first batch of projects funded by the Silk Road Fund has also been launched. We have quickened our pace in building industrial zones, cross-border economic cooperation zones, and port industrial zones with neighboring countries, as part of our efforts to promote cooperation on production capacity. Coming this May, China will host a high-level forum for International Cooperation on the Belt and Road to strengthen the complementarity of the development strategies of various countries. Preparations are well underway for the Asian Financial Cooperation Association, which is expected to be launched soon.

Fourth, we have actively promoted dialogue and cooperation on security. Every year, we have been hosting the Xiangshan Forum and the World Peace Forum to promote candid discussions on defense and security. China supports the improvement of regional security architecture. We hosted the Fifth EAS Regional Security Architecture Workshop and will host a Track 2 seminar on this topic to encourage more discussions. To better communicate China's security vision and policy, China issued in January a white paper outlining its policy on Asia-Pacific security cooperation, the first of its kind. Over the last couple of years, China has proposed one third of the total cooperation initiatives at the ARF, giving a strong boost to the exchange and cooperation in relevant areas.

The Chinese people are striving to realize the Chinese dream of great national rejuvenation. China's development will bring to Asia more opportunities for cooperation. China will continue to work with other regional countries to take regional cooperation to a higher level, and open greater prospects for Asia's development. In this context, I would like to share the following thoughts with you.

First, we need to be guided by the vision of an Asian community of shared future. This vision builds on our past experience of regional cooperation. As the ultimate goal of regional integration in Asia, it promises vast space for regional cooperation. To translate this vision into reality, all Asian countries need to work together for an Asian community of shared interests and shared responsibility by planning, building and benefiting together.

Second, we need to maintain the Asian features of our cooperation in improving the cooperation model. We should maintain our Asian-style cooperation model, featuring respect for ASEAN's centrality in East Asia cooperation, emphasis on coordination of various regional mechanisms, focus on development cooperation, and commitment to open regionalism. At the same time, we should improve our cooperation model by drawing lessons from other regions. We must ensure that all sub-regional mechanisms run in harmony to set an example of regional cooperation.

Third, we need to advance both development and security cooperation. We should make development a priority and enhance cooperation in such key areas as trade and finance, infrastructure, energy and environmental protection, etc. This is important for greater economic integration and the building of an economic community for common and sustainable development. At the same time, we should confront challenges head-on, and enhance dialogue, exchanges and cooperation on security. This is a necessary step as we seek to progressively build a regional security architecture that suits regional reality and meets the needs of all parties.

Fourth, we need to expand the reach of benefits from regional cooperation. Regional integration in Asia should not benefit only a few, just as globalization should not be a game of the elite. We should put people first, make the cooperation programs more inclusive, results-oriented, and beneficial to more countries and more peoples. This way, our people will see real benefits of regional cooperation, and give their wholehearted support. It will also help us avert the negative experience of other regions.

Actively Build the China-ASEAN Community of Shared Future

Remarks at the 23rd ASEAN-China Senior Officials' Consultation
May 18, 2017

This year marks the golden jubilee of ASEAN, an occasion of joy and celebration. One and a half years ago, ASEAN announced in Kuala Lumpur the establishment of the ASEAN Community, including political-security, economic and socio-cultural communities, and published the ASEAN Communities Vision and Blueprints 2025. As we noticed that at the 30th ASEAN Summit held successfully in the Philippines late last month, it was stressed that ASEAN should partner for change, engage the world, and build an integrated, peaceful, stable and resilient ASEAN upon its 50th anniversary. China congratulates ASEAN on its remarkable journey in the past 50 years, and commends the grand vision it has put forward. I would like to take this opportunity to share with you some of my thoughts.

Over the past half century, ASEAN countries have rebuilt Southeast Asia from the ruins of the Second World War and the Cold war, and worked together to turn the "Baltic in Asia" into one of the most peaceful regions in the world. It has established the first ever sub-regional community in Asia and profoundly changed the geopolitical landscape in East Asia. ASEAN has worked to invigorate the regional economy by integrating into the global market. It has set up partnerships with a dozen countries, including major countries, and international and regional organizations, and built regional cooperation architecture with ASEAN at its center. As such, ASEAN has evolved into a major force underpinning peace, stability and integration of the region and more broadly, a multi-polar world.

The "Asian Miracle" ASEAN has created is attributable to both internal and external factors, and a lot can be learned from this process. Externally, ASEAN, in its first 25 years, was under the shadows of the Cold War, confronting the two superpowers. ASEAN Cooperation was confined to the political and security fields and it was difficult to make real progress. In the following 25 years, with the end of the Cold War and peaceful settlement of the Cambodian issue, countries in the region regained control of regional affairs. They made full use of the peace dividend to usher ASEAN into a "golden era" of development. As fully linked with ASEAN countries on land and at sea, China is actually part of Southeast Asia. We highly cherish the peace and stability in this region. As the biggest neighbor of ASEAN, China pursued mutual learning and common development with ASEAN in the process of reform and opening-up.

As we have seen, ASEAN has stayed committed internally to a development path that suits its own conditions and is conducive to its own development during the second 25 years. This path is underpinned by three principles: First, ASEAN unity. The key to ASEAN's fast development lies in a shared sense of community that focuses on cooperation and development. With that, ASEAN has set a fine example for countries with different social systems, different religions and at different levels of development to seek strength through unity. Second, the ASEAN way. ASEAN stands by the principles of consensus-building, accommodating each other's comfort levels, and non-interference in internal affairs of its member states. These principles have proven to be effective for promoting cooperation in East Asia, served peace, stability and prosperity well and laid a solid foundation for ASEAN to grow from strength to strength. Third, ASEAN centrality in the regional architecture. ASEAN has taken active steps to build regional cooperation frameworks, played a leadership role in East Asia, cooperation and promoted sound interaction among major powers in this region. This has in turn helped ASEAN to build its influence and play an irreplaceable role in regional affairs.

ASEAN leaders have now put the priority on community building to turn ASEAN into a fine example of regionalism and active player in international affairs. We believe that, building on the experience and success of the past 50 years, ASEAN will achieve greater progress by further tapping its potential and move steadily toward its goal of building the East Asia Economic Community in the coming decades. Having said that, it's fair to say that ASEAN also faces new challenges as it advances community building. How to take the ASEAN Communities to higher levels? How to narrow the internal development gap? How to properly manage its relations with dialogue partners, including the issues of military alliances among some of its members and major powers? Addressing these challenges requires solidarity and joint efforts in the years ahead. In this context, I would like to offer our expectations and best wishes for ASEAN in the following areas:

First, we look to ASEAN to set an example for regional integration. China supports a stronger ASEAN as laid out in the ASEAN Community Vision 2025. We support ASEAN in strengthening the three pillars of political security, economic development and social-cultural progress for the benefit of regional countries and peoples. Given the recent anti-globalization sentiments, ASEAN communities serving as examples of regional integration will instill positive energy into the globalization process. We will further strengthen complementarity between China's Belt and Road Initiative and ASEAN's Connectivity Master Plan. We are ready to contribute to ASEAN's community-building and regional integration efforts by promoting sub-regional cooperation mechanisms such as the Lancang-Mekong Cooperation and BIMP-EAGA.

Second, we look to ASEAN to lead the regional cooperation process. ASEAN has been a source of regional stability and development. We hope ASEAN will further tap its central role in regional cooperation. Given the current context, we look to ASEAN to uphold the banner of globalization and regional integration and contribute even more to the sound and sustainable development and prosperity of East Asia. China supports ASEAN in continuing its role as "architect" and "driver" of regional cooperation. We encourage ASEAN to speed up the RECP negotiations, press ahead with the East Asia Economic Community, and explore with other parties the forward-looking regional security architecture, which reflects regional realities and meets the needs of all parties. Such efforts will be conducive to the sound and steady development momentum in East Asia.

Third, we look to ASEAN to serve as an anchor for regional peace and stability. A peaceful and stable regional environment is in the best interest of all ASEAN members and China. Over the years, all of your countries have worked closely to fully and effectively implement the DOC and safeguard peace and stability in the South China Sea. This has created a sound environment for the development of our region. In recent months, we worked together to bring down tension over the South China Sea, and made good progress in DOC implementation and COC consultation. We should cherish these hard-won gains. We hope and we believe that ASEAN friends will continue to focus on the big picture, properly handle sensitive issues related to the disputes, and uphold regional peace and development. In this respect, I would like to say that the issue of South China Sea is not an issue between ASEAN and China, but an issue among some coastal states. China is also prepared to initiate cooperation process among states at an appropriate time.

Fourth, we look to ASEAN to be a constructive player in the international arena. China supports ASEAN's policy of friendship and cooperation with all its dialogue partners. Such a policy will serve ASEAN well in better integrating into the region and embracing the world. We look forward to ASEAN 's continued positive role in facilitating interactions among other major countries by leveraging the East Asia cooperation frameworks, and work with them to advance regional development. We encourage ASEAN to play a constructive role on a broader international stage. For example, you are welcome to join forces with China and other dialogue partners on discussing major global issues such as economic globalization, climate change and the implementation of the 2030 Agenda for Sustainable Development.

Since the inception of our dialogue relations in 1991, China and ASEAN have seen deepening political mutual trust, growing practical cooperation, and joint responses to the Asian financial crisis. Our fast-growing, comprehensive and productive relationship has catalyzed ASEAN's relations with other dialogue partners, and set a good example for East Asia regional cooperation. Next year will mark the 15th anniversary of China-ASEAN strategic partnership. It will be a crucial moment to review the past and deliver a better relationship for the future. For China, ASEAN will always be a priority in our neighborhood policy. We are ready to work with ASEAN to further enhance mutual trust, deepen cooperation, upgrade China-ASEAN strategic partnership, and move toward a closer community of shared future.

Advance ASEAN Plus Three Cooperation to a New Stage

Keynote speech at the opening ceremony of the 15th East Asia Forum
June 30, 2017

This year marks the 20th anniversary of 10+3 cooperation. 20 years ago, with the strong momentum of globalization, regional integration and multi-polarity, countries in East Asia launched the 10+3 cooperation process in a joint fight against the Asian financial crisis. Since then, this framework has weathered two large financial crises, opened more than 20 cooperation areas and established over 60 dialogue and cooperation mechanisms. It is regarded as the main channel of East Asia cooperation and one of the most comprehensive and effective cooperation frameworks in Asia, making important contribution to dialogue and cooperation among regional countries and to regional development and prosperity. The regional cooperation has made positive progress. In the process of 10+3 cooperation, regional countries have cultivated an East Asian cooperation culture, featuring three words that start with the letter c:

The first word is **consensus**. All member states support the centrality of ASEAN in East Asia cooperation, adhere to the principles of mutual respect, mutual benefit and consensus, accommodating the comfort levels of all parties, and keep broadening areas and deepening levels of cooperation. All sides have actively boosted common and sustainable development while speeding up their own development, actively promoted common and cooperative security while seeking their own security, discussed and took joint actions in response to common challenges while pushing forward dialogues and cooperation in various areas. All this has further enhanced the status of 10+3 cooperation.

The second word is **connectivity**, which is crucial for deepening cooperation. We have been pushing forward physical connectivity, supporting ASEAN in the implementation of the MPAC 2025, encouraging regional countries to speed up the building of infrastructures, including airports, ports, roads, bridges and facilities on energy and telecommunication. We have been pushing forward economic and financial connectivity, speeding up the negotiations on RCEP and China-Japan-ROK FTA, and working on the building of a regional financial safety net with CMIM at its core. We have been pushing forward people-to-people connectivity by establishing East Asia Forum, East Asia Business Council, NEAT and other dialogue mechanisms, and conducting programs such as exchanges of cultural cities, official visits and exchanges of universities. We have been pushing forward connectivity of cooperation frameworks, increasing the complementarity and collaboration among the 10+3, other ASEAN-led frameworks and China-Japan-ROK cooperation. Promoting connectivity is an important part of the Belt and Road Initiative. All 10+3 members sent high-level representatives to the Belt and Road Forum for International Cooperation held in Beijing in May. The parties held in-depth discussions on regional connectivity and signed a great many cooperation agreements, injecting new vigor to East Asian connectivity.

The third word is **comprehensiveness**. We have expanded the scope of cooperation from traditional priorities of finance, food, energy, disaster management and poverty reduction to 24 areas of political and security, economy and finance, sustainable development and social and culture, and established a highly comprehensive cooperation mechanism. In recent years, regional foreign exchange reserve pool of CMIM has been expanded to US$240 billion. AMRO has been upgraded to an international organization. Credit Guarantee and Investment Facility (CGIF) under Asia Bond Market Initiative has operated soundly. The 10+3 Emergency Rice Reserve (APTERR) has started aid to disaster-affected area. Negotiation on RCEP is accelerating. The 10+3 has also increased attention to non-traditional security issues and actively conducted dialogue and cooperation in areas including disaster relief and control of infectious diseases.

10+3 cooperation has come a long way and made great achievements in the past 20 years. Since we are now in Chairman Mao Zedong's hometown, I want to quote one of his famous lines, that is, "Nothing is impossible if you work hard enough." This is a bold statement full of positive energy and enhanced our confidence in strengthening the 10+3 cooperation. Standing at a new starting point, we should aim high and make big strides to bring 10+3 cooperation to a higher level and open a better future for East Asia cooperation. To this end, I want to share with you some of my thoughts.

First, strengthen overall planning and institution building. 10+3 cooperation should be tailored to the regional situation and the needs of regional countries, which should guide the formulation of the work plans in a systematic way. We should exert the role of CPR+3, make full use of the 10+3 cooperation fund, encourage East Asia Forum and NEAT to provide constructive suggestions, and facilitate the setting-up of a 10+3 Unit in the ASEAN Secretariat. We should promote balance, coordination and synergy between 10+3 and other ASEAN-led frameworks to jointly contribute to East Asia integration. We should push forward a healthy, stable and sustainable China-Japan-ROK cooperation to reinforce 10+3 cooperation.

Second, set specific goals and approaches and work towards a shared vision. The East Asia Economic Community is an important target of 10+3 cooperation. It is also the only way towards the establishment of an East Asia Community and Asia community with a shared future. We should keep enhancing economic and trade cooperation, and move towards the EAEC by focusing on connectivity, free trade and financial cooperation. We should take into account the different development levels and needs of regional countries, and pursue an economic community with East Asian features by tackling easier tasks first and setting reasonable goals. The experiences of the EU and ASEAN show that drawing a road map is an essential step for establishing an economic community. During the 10+3 Leaders' Meeting last year, Premier Li Keqiang of China proposed dialogues on the drafting of an EAEC blueprint. I hope that this forum could help an early drafting of the blueprint.

Third, deepen practical cooperation and provide enduring driving force. In the financial area, we should enhance the effectiveness of CMIM, conduct periodical evaluation, and explore the possibility of local currency contribution. We should support AMRO's bid to become a permanent observer of the UN General Assembly and build an authoritative monitoring platform of the regional economy. In the area of food security, we should support enhancing the 10+3 APTERR for it to play a bigger role in stabilizing rice market and balancing supply and demand. In the area of free trade, we should endeavor to bring about early conclusion of RCEP and CJK FTA negotiations to contribute to the building of an Asia-Pacific Free Trade Area. In the area of sustainable development, we should enhance cooperation in poverty reduction, SME development, tourism and cultural exchanges in line with the spirit of "10+3 Statement on Promoting Sustainable Development" and push forward more effective implementation of the UN 2030 Sustainable Development Agenda in East Asia.

Fourth, create highlights of cooperation and inject new vitality. Since the low-hanging fruits are almost gone, we need to explore new directions for 10+3 cooperation. Given the complementarity of industrial structures of East Asian countries, China, Japan and the ROK should leverage their strength in equipment manufacturing to conduct industrial capacity cooperation with ASEAN countries. The goal should be to innovate industrial cooperation model and push forward the upgrading of the East Asia economy. SMEs have great potential in promoting regional economic development. Exchange and cooperation among them should be encouraged to push forward innovation and create employment. There is a great demand for infrastructure development in East Asia. Relevant countries should make full use of finance, technology and expertise to help countries speed up infrastructure development and promote regional connectivity and integration.

East Asia Cooperation and International Rule of Law

Keynote speech at the 2017 Colloquium on International Law: Common future in Asia
Hong Kong, July 7, 2017

Seventy-two years ago, the modern international legal system was established on the basis of the United Nations Charter after the end of the Second World War. Chapter eight of the Charter sets forth special provisions on regional arrangements. This has made regional cooperation frameworks an inherent part of the international system and global governance and an important platform for the promotion of international law. The end of the Cold War has seen regional cooperation flourish on all continents. East Asia has become a star-performer in global development and a region with the greatest vitality and potential. Why East Asia cooperation can come this far and how to maintain its good momentum merits some thinking.

Looking back to history, regional cooperation in East Asia has gone through roughly two stages.

The first stage was the 40 plus years from the end of the Second World War to the end of the Cold War. After the end of the Second World War, East Asian countries gained independence from colonial rule through hard struggle, and gradually joined the international system with the United Nations at the core as equal members of the international community. However, the Cold War confrontation between the two military blocs had all but ripped the region apart, leading to prolonged large-scale wars in Northeast Asia and Southeast Asia. Born in 1967, ASEAN lived under the shadows of the Cold War for much of its early years when cooperation was limited to political and security fields. Towards the end of the Cold War, some East Asian countries seized the opportunity to achieve fast economic development. Japan's economy took off. Four East Asian Tigers arose. China, Malaysia, Thailand and other countries began to vigorously develop an export-oriented economy.

The second stage of regional cooperation in East Asia was the 20 plus years since the end of the Cold War in 1991, which greatly invigorated East Asia cooperation. Driven by the wave of economic globalization and regional integration, East Asian countries achieved a collective rise, with booming regional cooperation. ASEAN grew rapidly, expanded its members from six maritime states to ten countries in Southeast Asia and established the ASEAN Free Trade Area. Cultivating a circle of ten dialogue partners covering major global countries, ASEAN became the driver of East Asia cooperation. Aware of the importance of multilateral security dialogues, East Asian countries set up the ASEAN Regional Forum after the Cold War. The outbreak of the Asian financial crisis in 1997 and weakened confidence in international financial governance led to the setting up of regional frameworks between ASEAN and China-Japan-ROK (so-called 10+3) and ASEAN's 10+1 frameworks with China, Japan and the ROK respectively. East Asia Summit as the strategic forum led by leaders was established, bringing the United States and Russia into the fold. This marked the formation of the East Asia cooperation architecture.

An important lesson we have learned from the history of East Asia development over the past 70 years is this: Only through unity, hard work and cooperation can we achieve the revitalization and development of East Asia and ensure regional peace and prosperity. This has also become an internal driving force for regional cooperation in East Asia.

Since the beginning of the new century, regional cooperation in East Asia has entered a fast lane, and made remarkable progress, which has effectively promoted regional integration in Asia.

First, the cooperation architecture has been further enriched. East Asia cooperation with the centrality of ASEAN has been enhanced by the addition of ADMM Plus, Expanded ASEAN Maritime Forum and other fresh platforms. The ASEAN community was founded in 2015. This year marks the 50th anniversary of ASEAN. Under the 10+3 Chiang Mai Initiative Multilateralization (CMIM), a foreign exchange reserve pool of US$240 billion was set up. ASEAN+3 Macroeconomic Research Institute (AMRO) was upgraded to an international organization. China-Japan-ROK cooperation has established 21 ministerial level meetings and more than 60 governmental consultation mechanisms.

Second, sub-regional cooperation has been gathering momentum. The Lancang-Mekong Cooperation took off with a pragmatic and highly-efficient LMC speed. BIMP-EAGA is speeding up cooperation on a transport network connection. The BCIM Economic Corridor is advancing steadily with three meetings of joint study groups held and the synthesis report close to completion. The Greater Tumen Initiative is conducting close cooperation in tourism, energy and other fields, and moving towards an independent international organization.

Third, fruitful results have been achieved in cooperation. Trade liberalization and facilitation has gained more support. Negotiations on RCEP and China-Japan-ROK Free Trade Area are moving forward at a faster pace. Regional connectivity has been strengthened with the construction of major infrastructure projects, including the China-Pakistan Economic Corridor, Jakarta-Bandung High-speed Rail and China-Laos Railway. Dialogues and cooperation in non-traditional security issues such as disaster relief, counter-terrorism and transnational crimes are deepening. Joint exercises and law-enforcement operations have been carried out.

Having said all this, we should not fail to see the many constraints facing East Asia cooperation. First, East Asia lacks a framework that covers the whole region and all areas of cooperation. Second, the building of a security cooperation framework is relatively lagging behind, becoming the short plank in East Asia cooperation. Third, intense interactions among major countries can sometimes negatively impact East Asia cooperation.

China has been a strong supporter, active participant and chief mover of East Asia cooperation. In the past five years, under the visionary leadership of the Party Central Committee with General Secretary Xi Jinping at the core, China has made indispensable contribution to expanding, enriching and upgrading East Asia cooperation.

First, we contributed a Chinese vision on cooperation. Since President Xi Jinping proposed the building of a community of shared future at the Boao Forum for Asia Annual Conference 2013, we have sought to implement this vision in the regional context by building a China-ASEAN, an Asian, and Asia-Pacific community of shared future. These concepts point to the ultimate goal for East Asia cooperation. China also proposed the concept of innovative, coordinated, green, open and shared development, promoted an Asia-Pacific partnership based on mutual trust, inclusiveness and win-win cooperation, and advocated an Asian Security Concept featuring common, comprehensive, cooperative and sustainable security. These visions show the way forward for regional economic and social development and security cooperation.

Second, we actively pushed forward practical cooperation. China has advanced the Belt and Road Initiative to facilitate the common development of countries along the routes. We promoted cooperation in industrial capacity and vigorously pursued the construction of industrial parks, cross-border economic cooperation zones and port-side industrial parks in East Asian countries. This May, China successfully held the Belt and Road Forum on International Cooperation, which enhanced the complementarity of development strategies with relevant countries. China has also actively promoted liberalization of regional trade and investment, signed FTA with the ROK and Australia and completed negotiations on upgrading China-ASEAN FTA to support the building of Asia-Pacific FTA.

Third, we have supported sub-regional cooperation mechanisms. China and the five countries along the Mekong river together created the Lancang-Mekong Cooperation mechanism, which has injected new impetus for the ASEAN Community building and the regional integration process. The Asian Infrastructure Investment Bank (AIIB) proposed by China has been formally launched, providing a new channel for regional countries to overcome the bottleneck of infrastructure financing. As a new financing platform for the Belt and Road Initiative, the Silk Road Fund financed by China has approved 15 projects with a total investment of US$6 billion.

Fourth, we have actively promoted security dialogues and cooperation. In January this year, we issued the first White Paper on China's Asia-Pacific Security Cooperation Policy, which voiced our strong will to enhance regional security cooperation. We hold the Xiangshan Forum and World Peace Forum annually to enhance exchanges in defense and security areas. We support improving regional security architecture and have been actively promoting relevant discussions under the EAS framework. China and ASEAN launched Defense Ministers' Meeting, and established Ministerial Dialogue on Law Enforcement Security Cooperation. The two sides are comprehensively and effectively implementing the DOC, and making positive progress on maritime cooperation. The negotiation on COC has been moving forward steadily. The draft framework of COC was completed during our meetings in Guiyang in May.

Fifth, we have actively facilitated people-to-people and cultural exchanges. China has become the biggest source of overseas tourists of ASEAN. China and ASEAN countries are connected every day by over 1,000 flights. In 2016, people-to-people exchanges between the two sides surpassed 30 million, and student exchanges reached nearly 200,000. China and ASEAN designated 2016 and 2017 as Year of Educational Exchange and Year of Tourism Cooperation respectively. We will pursue people-to-people and cultural exchanges as new pillars of China-ASEAN relations. There have been dynamic exchanges between China, Japan and the ROK, with visits of over 20 million each year and moving towards the goal of 30 million in 2020. China has actively implemented projects such as Asia Campus and Cultural Cities of East Asia to push forward educational and cultural exchanges among the three countries.

The international and regional landscape has been undergoing profound changes. The world is confronted with a host of challenges, from lingering effect of the international financial crisis, rising anti-globalization sentiments and trade protectionism, to flare-up of hotspot issues and growing non-traditional threats. Enhancing regional cooperation has never been more important. East Asia cooperation now stands at a new starting point. The ardent expectation of regional countries is upon us, and we are ready to shoulder greater tasks of safeguarding regional peace and prosperity.

As one of the important members of the East Asia family, China will continue to expand its participation in and engagement with East Asia and make joint efforts with all parties to build a lasting stable and prosperous East Asia. Let me take this opportunity to share some thoughts on the future development of East Asia cooperation.

First, we should hold high the banner of East Asia community and set up long-term goals. Our primary goal now is to build the East Asia Economic Community, promote trade, investment and financial cooperation, and deepen regional economic integration. We are ready to work with other countries to build open, integrated, balanced and win-win economic cooperation and lead economic globalization and trade liberalization process.

Second, we should uphold the spirit of East Asia cooperation and build the inclusive cooperation architecture. Regional countries evolved the ASEAN Way featuring ASEAN centrality, mutual respect, consensus and accommodating the comfort levels of all parties, and cultivated the concept and habit of open, inclusive, and win-win cooperation. This is a valuable asset of East Asian countries in getting along well and properly handling their differences. We should carry forward this cooperation spirit with East Asian features, adhere to open regionalism, and encourage countries outside the region to play a positive and constructive role in promoting East Asia cooperation, which can complement and reinforce Asia-Pacific cooperation.

Third, we should promote inclusive development to achieve win-win outcomes. Expanding overlapping interests is an important basis for developing relations among countries and also a fundamental guarantee for regional peace and stability. With the weak recovery of global economy and the downward economic pressure, we should focus on development cooperation to narrow development gaps, and speed up restructuring to achieve innovative, coordinated, green, open and shared development.

Fourth, we should build a new framework for security cooperation to make up the short plank in security. East Asia is faced with complicated security challenges, which call for the regional security architecture that reflects the regional reality and the demands of all sides. We should be guided by common, comprehensive, cooperative and sustainable security to gradually build the future-oriented regional security cooperation architecture in Asia.

Author's Note

I am grateful to the National Institute for South China Sea Studies for compiling and publishing this collection of my essays, articles and speeches in recent years on Asian regional cooperation.

It has been over a year since I left my post as Chinese Vice Minister for Foreign Affairs in July 2017 and arrived in New York, serving as United Nations Under-Secretary-General for Economic and Social Affairs. It is a new and unique experience indeed to work at a different diplomatic stage known as the United Nations.

As the Under-Secretary-General for Economic and Social Affairs, I have a mandate to lead the UN Department of Economic and Social Affairs (UN DESA) in support and service to the UN General Assembly and Economic and Social Council (ECOSOC), promote global sustainable development in the economic, social and environmental fields, and implement the 2030 Agenda for Sustainable Development and the Sustainable Development Goals through consensus and capacity building.

As a participant in Asian regional cooperation for many years, I have been thinking about how Asia may contribute to global development through regional cooperation in East Asia.

Speaking of East Asian regional cooperation, I must mention the Lancang-Mekong Cooperation (LMC), a major sub-regional cooperation mechanism in East Asia. As China and the five Mekong countries have a shared river and work towards a shared future, it is only natural that the countries along the same river have created a cooperative mechanism of their own. Being developing countries, the six LMC countries face a common task of growing the economy and improving the people's livelihoods.

The LMC mechanism makes an excellent complement to China-ASEAN cooperation and is a new practice in South-South cooperation. It will help unlock the potentials of the development of the LMC countries and provide an opportunity for them to draw on each other's economic strength. It will also help boost economic and social development in the region and narrow the development gap, and contribute to the implementation of the UN 2030 Agenda for Sustainable Development.

I have had the pleasure and privilege of participating in the creation of the LMC and have been following its development with great interest. I am more than happy that Lancang-Mekong Cooperation is thriving and has made considerable progress since its official launch at the first LMC Leaders' Meeting in Sanya, China in March 2016. The cooperation is advancing steadily in the three pillar areas of political and security issues, economic growth and sustainable development, and social, cultural and people-to-people exchanges, and has produced tangible results. A large number of cooperation projects on the ground are underway. The LMC speed and efficiency are widely recognized and commended. The joint working groups on the six priority areas, namely, connectivity, industrial capacity, cross-border economic cooperation, water resources, agriculture and poverty reduction, are up and running. The ministries of foreign affairs of the LMC countries have set up their LMC national secretariats. A five-year LMC plan of action is in the making. In January 2018, the Second LMC Leaders' Meeting was held in Cambodia.

The 19th National Congress of the Communist Party of China (CPC) has reaffirmed the building of a community of shared future for mankind as a goal of the country. The LMC aims to promote the complementarity of national development strategies, coordinate cooperation resources, share fruits of development, and galvanize LMC countries in a joint effort to build a Lancang-Mekong community of shared future, one that features unity, mutual assistance, consultation on an equal footing, and mutually beneficial cooperation. The LMC has become a hallmark of the envisaged Asian and global community of shared future.

I am confident that with the joint efforts of its participating countries, the LMC will have a multiplier effect in promoting prosperity in Asia and delivering practical benefits to the people living in the region. It is also my belief that the "LMC approach" will contribute to sustainable economic and social progress in the world.

Economic globalization is facing strong headwinds never seen before. Asia's development and China's rising influence are the focus of the world's attention. East Asian nations are expected to play a greater role in promoting economic globalization. The future of East Asia lies in stronger and closer Asian regional cooperation. What China can and should do is, in my view, to embrace the Asia-Pacific and the world while focusing on East Asia, and work actively and prudently to promote Asian regional cooperation.

Liu Zhenmin
New York, August 2018

Liu Zhenmin

Starting from July 26, 2017, Mr. Liu serves as United Nations Under-Secretary-General for Economic and Social Affairs.

Mr. Liu was born in August 1955 in Shanxi Province, China. He studied successively in the Department of Western Language and Literature and the Department of Law of Peking University, and graduated in 1981 with a master's degree in law. He joined the Ministry of Foreign Affairs of the People's Republic of China in early 1982, and started his 35-year diplomatic career during which he held many important posts including China's Vice Minister of Foreign Affairs, Assistant Minister of Foreign Affairs, Ambassador Extraordinary & Plenipotentiary and Chinese Permanent Representative to the United Nations Office at Geneva and Other International Organizations in Switzerland, Deputy Permanent Representative of China to the United Nations and Ambassador Extraordinary and Plenipotentiary, Director-General and Deputy Director-General of the Department of Treaty and Law of the Ministry of Foreign Affairs. During the six years and more as Assistant Minister and Vice Minister of Foreign Affairs, he was in charge of Asian, treaty and law, border and ocean affairs, playing an important role in promoting regional peace and stability, mutual trust and cooperation among Asian countries, regional integration and innovation of cooperation mechanisms.

Mr. Liu has long been devoted to promoting the development of and cooperation on global affairs, and is renowned in the international community. He has been deeply involved in climate change negotiations for ten years, including negotiations on the Kyoto Protocol and the Paris Agreement, as well as important international activities to protect our planet, including those related to Antarctic issues and maritime affairs. He currently also serves as a member of the Permanent Court of Arbitration and an alternate member of the Institute of International Law.

At present, Liu is responsible for leading the United Nations Secretariat support for the follow-up of the implementation of the 2030

Agenda for Sustainable Development. He is also in charge of leading the policy analysis and capacity-building of the United Nations on many economic, social and environmental issues, such as sustainable development, internet governance, climate change, financing for development, youth and aging, and supporting intergovernmental negotiations in relevant fields.